AF804500

857
Me
2e

Nonlinear Systems and Complexity

Volume 31

Series editor

Albert C. J. Luo
Southern Illinois University
Edwardsville, IL, USA

Nonlinear Systems and Complexity provides a place to systematically summarize recent developments, applications, and overall advance in all aspects of nonlinearity, chaos, and complexity as part of the established research literature, beyond the novel and recent findings published in primary journals. The aims of the book series are to publish theories and techniques in nonlinear systems and complexity; stimulate more research interest on nonlinearity, synchronization, and complexity in nonlinear science; and fast-scatter the new knowledge to scientists, engineers, and students in the corresponding fields. Books in this series will focus on the recent developments, findings and progress on theories, principles, methodology, computational techniques in nonlinear systems and mathematics with engineering applications. The Series establishes highly relevant monographs on wide ranging topics covering fundamental advances and new applications in the field. Topical areas include, but are not limited to: Nonlinear dynamics Complexity, nonlinearity, and chaos Computational methods for nonlinear systems Stability, bifurcation, chaos and fractals in engineering Nonlinear chemical and biological phenomena Fractional dynamics and applications Discontinuity, synchronization and control.

More information about this series at http://www.springer.com/series/11433

J. A. Tenreiro Machado • Necati Özdemir
Dumitru Baleanu
Editors

Numerical Solutions of Realistic Nonlinear Phenomena

Springer

Mängelexemplar

Editors
J. A. Tenreiro Machado
Institute of Engineering
Polytechnic of Porto
Porto, Portugal

Necati Özdemir
Balıkesir University
Department of Mathematics
Balıkesir, Turkey

Dumitru Baleanu
Department of Mathematics
& Computer Science
Çankaya University
Ankara, Turkey

ISSN 2195-9994 ISSN 2196-0003 (electronic)
Nonlinear Systems and Complexity
ISBN 978-3-030-37140-1 ISBN 978-3-030-37141-8 (eBook)
https://doi.org/10.1007/978-3-030-37141-8

© Springer Nature Switzerland AG 2020
This work is subject to copyright. All rights are reserved by the Publisher, whether the whole or part of the material is concerned, specifically the rights of translation, reprinting, reuse of illustrations, recitation, broadcasting, reproduction on microfilms or in any other physical way, and transmission or information storage and retrieval, electronic adaptation, computer software, or by similar or dissimilar methodology now known or hereafter developed.
The use of general descriptive names, registered names, trademarks, service marks, etc. in this publication does not imply, even in the absence of a specific statement, that such names are exempt from the relevant protective laws and regulations and therefore free for general use.
The publisher, the authors, and the editors are safe to assume that the advice and information in this book are believed to be true and accurate at the date of publication. Neither the publisher nor the authors or the editors give a warranty, expressed or implied, with respect to the material contained herein or for any errors or omissions that may have been made. The publisher remains neutral with regard to jurisdictional claims in published maps and institutional affiliations.

This Springer imprint is published by the registered company Springer Nature Switzerland AG.
The registered company address is: Gewerbestrasse 11, 6330 Cham, Switzerland

Preface

The International Conference on Applied Mathematics in Engineering (ICAME'18) was successfully held in the period of 27–29 June 2018 in Burhaniye, Turkey. The conference provided an ideal academic platform for researchers to present the latest research and evolving findings of applied mathematics on engineering, physics, chemistry, biology and statistics. During the conference,

- Three plenary lectures (by Prof. Dr. Albert Luo, Prof. Dr. J. A. Tenreiro Machado and Prof. Dr. Jordan Hristov under the chairship of Prof. Dr. Dumitru Baleanu)
- Three invited talks (by Prof. Dr. Carla Pinto, Prof. Dr. Mehmet Kemal Leblebicioglu and Prof. Dr. Ekrem Savas)
- A total of 224 oral presentations (in eight parallel sessions)

have been successfully presented by participants from 15 different countries, i.e. Algeria, Argentina, Bulgaria, Libya, Germany, India, Morocco, Nigeria, Portugal, Saudi Arabia, South Africa, Turkey, the United Arab Emirates, the United Kingdom and the United States of America.

The members of the organizing committee were Ramazan Yaman (Turkey), J. A. Tenreiro Machado (Portugal), Necati Ozdemir (Turkey) and Dumitru Baleanu (Romania, Turkey).

We would like to thank all the members of the Scientific Committee for their valuable contribution forming the scientific face of the conference, as well as for their help in reviewing the contributed papers. We are also grateful to the staff involved in the local organization.

This work organized in two volumes publishes a selection of extended papers presented at ICAME'18 after a rigorous peer-reviewing process. The first volume of the book *Numerical Solutions of Realistic Nonlinear Phenomena* contains 12 high-quality contributions.

This collection covers new aspects of numerical methods in applied mathematics, engineering, and health sciences. It provides recent theoretical developments and new techniques based on optimization theory, partial differential equations (PDEs), mathematical modelling and fractional calculus that can be used to model and understand complex behaviour in natural phenomena. Specific topics covered in detail include new numerical methods for nonlinear PDEs, global optimization, unconstrained optimization, detection of HIV-1 protease, modelling with new fractional operators, analysis of biological models and stochastic modelling.

We thank all the referees and other colleagues who helped in preparing this book for publication. Finally, our special thanks are due to Albert Luo and Michael Luby from Springer for their continuous help and work in connection with this book.

Porto, Portugal J. A. Tenreiro Machado
Balıkesir, Turkey Necati Özdemir
Ankara, Turkey Dumitru Baleanu

Contents

Chapter 1
Monotone Iterative Technique for Non-autonomous Semilinear Differential Equations with Non-instantaneous Impulses

Arshi Meraj and Dwijendra Narain Pandey

1.1 Introduction

To describe abrupt changes, for example, harvesting, natural disasters and shocks, the differential equations having instantaneous impulses are used. The theory of instantaneous impulsive equations has various applications in mechanics, control, electrical engineering, medical and biological fields. One may see [1, 3, 7, 22, 23] for more details.

In pharmacotherapy, certain dynamics of evolution process could not be explained by the models having instantaneous impulses. For example, the introduction of the drugs in bloodstream and its absorption to the body are continuous and gradual process, the above situations can be interpreted as impulsive actions which start abruptly and stay active on a finite time interval. Hernández [17] initially considered Cauchy problems of first order non-instantaneous impulsive evolution equations. Fractional non-instantaneous impulsive differential system is considered by Kumar et al. [21] to establish the existence and uniqueness of mild solutions. Chen et al. [10] investigated the mild solutions for first order evolution equations having non-instantaneous impulses using noncompact semigroup. Kumar et al. [20] derived a set of sufficient conditions for the existence and uniqueness of mild solutions to a fractional integro-differential equation with non-instantaneous impulses.

Monotone iterative technique (in short MIT) is a useful method for the study of existence and uniqueness of mild solutions. In MIT, we construct monotone sequences of approximate solutions converging to extremal mild solutions. This technique is first used by Du [13] to find extremal mild solutions for a differential equation. The results are extended for nonlocal differential equations by Chen [8].

A. Meraj (✉) · D. N. Pandey
Indian Institute of Technology Roorkee, Roorkee, Uttarakhand, India

© Springer Nature Switzerland AG 2020

J. A. T. Machado et al. (eds.), *Numerical Solutions of Realistic Nonlinear Phenomena*, Nonlinear Systems and Complexity 31, https://doi.org/10.1007/978-3-030-37141-8_1

Further the technique is used for an impulsive integro-differential equation by Chen and Mu [9]. Mu [25] first applied the MIT for fractional evolution equations. Mu and Li [27] generalized the results for impulsive fractional differential equations by using MIT. Later on, the result has been extended for nonlocal condition by Mu [26]. Kamaljeet [18] and Renu [6] applied MIT for fractional differential equations having finite and infinite delay, respectively. Evolution equations having non-instantaneous impulses are studied by Chen et al. [32] via MIT. In [24], authors first applied MIT for non-autonomous differential system.

Non-autonomous nonlocal integro-differential equations are studied by Yan [30] with the help of evolution system, Schauder and Banach fixed point theorems. Non-autonomous differential system having deviated arguments is considered by Haloi et al. [15] to study the existence and asymptotic stability via analytic semigroup and Banach contraction theorem. In [5], authors generalized the results of [15] for instantaneous impulsive non-autonomous differential equations having iterated deviated arguments.

In literature, no work yet available for non-autonomous differential system having non-instantaneous impulses by using MIT. Inspired by this, we consider the following non-autonomous system having non-instantaneous impulsive condition in a partially ordered Banach space X:

$$\begin{aligned} x'(t) + \mathbf{A}(t)x(t) &= \mathscr{F}(t, x(t)), \quad t \in \cup_{i=0}^{m}(s_i, t_{i+1}], \\ x(t) &= \gamma_i(t, x(t)), \quad t \in \cup_{i=1}^{m}(t_i, s_i], \\ x(0) &= x_0, \end{aligned} \tag{1.1}$$

where $\{\mathbf{A}(t) : t \in J = [0, b]\}$ is a family of linear closed operators. The nonlinear function $\mathscr{F} : J \times X \to X$ and non-instantaneous impulsive functions $\gamma_i : (t_i, s_i] \times X \to X$, $i = 1, 2, \ldots, m$ are suitable functions, $0 < t_1 < t_2 < \ldots < t_m < t_{m+1} := b$, $s_0 := 0$ and $s_i \in (t_i, t_{i+1})$ for each $i = 1, 2, \ldots, m$ and $x_0 \in X$.

The article is arranged as follows: Section 1.2 is related with some basic theory. The existence and uniqueness of extremal mild solutions for the system (1.1), nonlocal problem and integro-differential system corresponding to (1.1) are established in Sects. 1.3, 1.4 and 1.5. In last section, we present an example in the favour of our results.

1.2 Preliminaries

Suppose that $(X, \|\cdot\|, \leq)$ is a partially ordered Banach space, $\mathscr{C}(J, X)$ denotes the space of all continuous maps from J to X, endowed with supremum norm. Consider $\mathscr{PC}(J, X) = \{x : J \to X : x$ is continuous at $t \neq t_k$, $x(t_{k-}) = x(t_k)$ and $x(t_{k+})$ exists for all $k = 1, 2, \ldots .m\}$, endowed with supremum norm, $\mathscr{P} = \{y \in X : y \geq 0\}$ (0 is the zero element of X) denotes the positive cone of X.

The positive cone is known as normal if we have a real number $\mathcal{N} > 0$ satisfying $0 \leq x_1 \leq x_2 \Rightarrow \|x_1\| \leq \mathcal{N}\|x_2\|$, for all $x_1, x_2 \in X$, the least value of $\mathcal{N}$ is named as normal constant. For $u, w \in \mathscr{PC}(J, X)$ with $u \leq w$, we will use the notation $[u, w] := \{y \in \mathscr{PC}(J, X) : u \leq y \leq w\}$ for an interval in $\mathscr{PC}(J, X)$, while $[u(t), w(t)] := \{z \in X : u(t) \leq z \leq w(t)\}(t \in J)$ for an interval in X, $\mathscr{L}^p(J, X)(1 \leq p < \infty)$ denotes the Banach space with norm $\|x\|_{\mathscr{L}^p(J,X)} = (\int_0^b \|x(t)\|^p dt)^{\frac{1}{p}}$.

First, we recall the definition and some basic properties of evolution system. We refer [14] and [28] for more details.

Definition 1.1 ([28]) Evolution system is a two parameter family of bounded linear operators $\mathscr{S}(t_1, t_2), 0 \leq t_2 \leq t_1 \leq b$ on a Banach space X satisfying:

1. $\mathscr{S}(s, s) = I$ (the identity operator).
2. $\mathscr{S}(t_1, t_2)\mathscr{S}(t_2, t_3) = \mathscr{S}(t_1, t_3)$ for $0 \leq t_3 \leq t_2 \leq t_1 \leq b$.
3. $(t_1, t_2) \to \mathscr{S}(t_1, t_2)$ is strongly continuous for $0 \leq t_2 \leq t_1 \leq b$.

The following assumptions are imposed on the family of linear operators $\{\mathbf{A}(t) : t \in J\}$ on X:

(A1) $\mathbf{A}(t)$ is closed densely defined operator, the domain of $\mathbf{A}(t)$ does not depend on t.
(A2) For $Re(\vartheta) \leq 0,\ t \in J$, the resolvent of $\mathbf{A}(t)$ exists and satisfies $\|\mathscr{R}(\vartheta; t)\| \leq \frac{\varsigma}{|\vartheta|+1}$, for some positive constant ς.
(A3) For some positive constants K and $\rho \in (0, 1]$, we have

$$\|[\mathbf{A}(\tau_1) - \mathbf{A}(\tau_2)]\mathbf{A}^{-1}(\tau_3)\| \leq K|\tau_1 - \tau_2|^{\rho}, \text{ for any} \tau_1, \tau_2, \tau_3 \in J.$$

Theorem 1.1 ([28]) *Suppose that the assumptions (A1)-(A3) hold, then $-\mathbf{A}(t)$ generates a unique evolution system $\{\mathscr{S}(t_1, t_2) : 0 \leq t_2 \leq t_1 \leq b\}$, which satisfies:*

(i) For some positive constant $\mathscr{M}$, we have $\|\mathscr{S}(t_1, t_2)\| \leq \mathscr{M}, 0 \leq t_2 \leq t_1 \leq b$.
(ii) For $0 \leq t_2 < t_1 \leq b$, the derivative $\frac{\partial \mathscr{S}(t_1,t_2)}{\partial t_1}$ exists in strong operator topology, is strongly continuous, and belongs to $B(X)$ (set of all bounded linear operators on X). Moreover,

$$\frac{\partial \mathscr{S}(t_1, t_2)}{\partial t_1} + \mathbf{A}(t_1)\mathscr{S}(t_1, t_2) = 0, \quad 0 \leq t_2 < t_1 \leq b.$$

Proposition 1.1 ([30]) *The family of operators $\{\mathscr{S}(t_1, t_2),\ t_2 < t_1\}$ is continuous in t_1 uniformly for t_2 with respect to operator norm.*

Definition 1.2 The evolution system $\mathscr{S}(t, s)$ is named as positive if it maps positive cone to the positive cone.

Theorem 1.2 ([28]) *If $\mathscr{F}$ satisfies uniform Hölder continuity on J with exponent $\alpha \in (0, 1]$ and the assumptions (A1)–(A3) hold, then the unique solution for the* *following linear problem*

$$x'(t) + \mathbf{A}(t)x(t) = \mathscr{F}(t), \quad 0 < t \leq b,$$
$$x(0) = x_0 \in X \tag{1.2}$$

is given by

$$x(t) = \mathscr{S}(t,0)x_0 + \int_0^t \mathscr{S}(t,\eta)\mathscr{F}(\eta)d\eta. \tag{1.3}$$

Definition 1.3 A mild solution to the problem (1.1) is a function $x \in \mathscr{PC}(J, X)$ satisfying

$$x(t) = \begin{cases} \mathscr{S}(t,0)x_0 + \int_0^t \mathscr{S}(t,\eta)\mathscr{F}(\eta, x(\eta))d\eta, & t \in [0, t_1], \\ \gamma_i(t, x(t)), & t \in \cup_{i=1}^m (t_i, s_i], \\ \mathscr{S}(t, s_i)\gamma_i(s_i, x(s_i)) + \int_{s_i}^t \mathscr{S}(t,\eta)\mathscr{F}(\eta, x(\eta))d\eta, & t \in \cup_{i=1}^m (s_i, t_{i+1}]. \end{cases} \tag{1.4}$$

Definition 1.4 A lower mild solution for the system (1.1) is $\omega_0 \in \mathscr{PC}(J, X)$ satisfying

$$\omega_0(t) \leq \begin{cases} \mathscr{S}(t,0)x_0 + \int_0^t \mathscr{S}(t,\eta)\mathscr{F}(\eta, \omega_0(\eta))d\eta, & t \in [0, t_1], \\ \gamma_i(t, \omega_0(t)), & t \in \cup_{i=1}^m (t_i, s_i], \\ \mathscr{S}(t, s_i)\gamma_i(s_i, \omega_0(s_i)) + \int_{s_i}^t \mathscr{S}(t,\eta)\mathscr{F}(\eta, \omega_0(\eta))d\eta, & t \in \cup_{i=1}^m (s_i, t_{i+1}]. \end{cases} \tag{1.5}$$

If the inequalities of (1.5) are opposite, solution is named as upper mild solution.

Now, let us recall the definition and some properties of Kuratowski measure of noncompactness.

Definition 1.5 Let $M(Y)$ be a family of bounded subsets of a Banach space Y, then the nonnegative function μ on $M(Y)$ defined as:

$$\mu(D) = \inf\{\varepsilon > 0 : D \subset \cup_{j=1}^n D_j, \ \text{diam}\,(D_j) < \varepsilon \ (j = 1, 2, \ldots, n \in N)\},$$

is called Kuratowski measure of noncompactness.

Lemma 1.1 ([11]) *Let X_1 and X_2 be complete norm spaces and $G, H \subset X_1$ are bounded, then we have:*

(i) *G is precompact if and only if $\mu(G) = 0$.*
(ii) $\mu(G \cup H) = \max\{\mu(G), \mu(H)\}$.
(iii) $\mu(G + H) \leq \mu(G) + \mu(H)$.
(iv) *If $f : dom(f) \subset X_1 \to X_2$ is Lipschitz continuous with Lipschitz constant C, then $\mu(f(S)) \leq C\mu(S)$, $S \subset dom(f)$ is bounded.*

Lemma 1.2 ([2]) *Let $G \subset \mathscr{C}(J,Y)$, where Y is a complete norm space, and $G(t) = \{f(t) : f \in G\}(t \in J)$. If G is equicontinuous and bounded in $\mathscr{C}(J,Y)$, then $\mu(G(t))$ is continuous on J and $\mu(G) = \max_{t\in J} \mu(G(t))$.*

Lemma 1.3 ([16]) *Let $\{f_n\} \subset \mathscr{C}(J,X)$ is a bounded sequence where X is complete norm space, then $\mu(\{f_n(t)\} \in \mathscr{L}^1(J,X)$, and*

$$\mu\left(\left\{\int_0^t f_n(\eta)d\eta\right\}_{n=1}^{\infty}\right) \leq 2\int_0^t \mu(\{f_n(\eta)\}_{n=1}^{\infty})d\eta.$$

Lemma 1.4 ([31]) *Let $c \geq 0$, $\beta > 0$, $y(t)$ and $z(t)$ are locally integrable nonnegative functions on $0 \leq t < T < +\infty$, such that*

$$z(t) \leq y(t) + c\int_0^t (t-s)^{\beta-1} z(s)ds,$$

then

$$z(t) \leq y(t) + \int_0^t \left[\sum_{n=1}^{\infty} \frac{(c\Gamma(\beta))^n}{\Gamma(n\beta)}(t-s)^{n\beta-1}y(s)\right]ds, \quad 0 \leq t < T.$$

1.3 Main Results

In this section, first we will discuss the existence of extremal mild solutions for (1.1), then the uniqueness will be discussed. Define $\mathscr{Q} : \mathscr{PC}(J,X) \to \mathscr{PC}(J,X)$ in the following way:

$$\mathscr{Q}x(t) = \begin{cases} \mathscr{S}(t,0)x_0 + \int_0^t \mathscr{S}(t,\eta)\mathscr{F}(\eta,x(\eta))d\eta, & t \in [0,t_1], \\ \gamma_i(t,x(t)), & t \in \cup_{i=1}^m (t_i,s_i], \\ \mathscr{S}(t,s_i)\gamma_i(s_i,x(s_i)) + \int_{s_i}^t \mathscr{S}(t,\eta)\mathscr{F}(\eta,x(\eta))d\eta, & t \in \cup_{i=1}^m (s_i,t_{i+1}]. \end{cases} \tag{1.6}$$

Note that $\mathscr{Q}$ is a well-defined map. We are interested to prove that the operator $\mathscr{Q}$ has a fixed point. Now, we state the assumptions required to prove the existence of extremal mild solutions:

(H0) X is a partially ordered Banach space with positive normal cone $\mathscr{P}$, $x_0 \in X$, and $-\mathbf{A}(t)$ generates a positive evolution system $\mathscr{S}(t,s)$ $(0 \leq s \leq t \leq b)$ on X.

(H1) The function $\mathscr{F} : \cup_{i=0}^m [s_i, t_{i+1}] \times X \to X$ is continuous, and satisfies

$$\mathscr{F}(t,y_1) \leq \mathscr{F}(t,y_2),$$

for $y_1, y_2 \in X$ and $\omega_0(t) \leq y_1 \leq y_2 \leq \nu_0(t)$.

(H2) The function $\gamma_i : \cup_{i=1}^{m}(t_i, s_i] \times X \to X$ is continuous, and satisfies

$$\gamma_i(t, y_1) \le \gamma_i(t, y_2),$$

for $y_1, y_2 \in X$ and $\omega_0(t) \le y_1 \le y_2 \le \nu_0(t)$.

(H3) For all $t \in \cup_{i=0}^{m}[s_i, t_{i+1}]$, and $\{y_n\} \subset [\omega_0(t), \nu_0(t)]$ a monotone increasing or decreasing sequences, we have

$$\mu(\{\mathscr{F}(t, y_n)\}) \le \mathscr{L}(\mu(\{y_n\})),$$

for some constant $\mathscr{L} > 0$.

(H4) There is a constant $\mathscr{L}_i > 0$ such that for all $t \in (t_i, s_i]$, $i = 1, 2, \ldots, m$, and $\{y_n\} \subset [\omega_0(t), \nu_0(t)]$ a monotone increasing or decreasing sequences, we have

$$\mu(\{\gamma_i(t, y_n)\}) \le \mathscr{L}_i \mu(\{y_n\}).$$

Theorem 1.3 *If the assumptions* $(A1)$–$(A3)$, $(H0)$–$(H4)$ *are satisfied, and* $\omega_0, \nu_0 \in \mathscr{PC}(J, X)$ *with* $\omega_0 \le \nu_0$ *are lower and upper mild solutions, respectively, for system* (1.1). *Then, there exist extremal mild solutions in the interval* $[\omega_0, \nu_0]$, *for the system* (1.1), *provided that* $\max_{i=1,2,\ldots,m} \{\mathscr{L}_i\} < 1$.

Proof Let us denote $I = [\omega_0, \nu_0]$. For any $x \in I$ and $\varrho \in [s_i, t_{i+1}]$; $i = 0, 1, 2, \ldots, m$, $(H1)$ implies

$$\mathscr{F}(\varrho, \omega_0(\varrho)) \le \mathscr{F}(\varrho, x(\varrho)) \le \mathscr{F}(\varrho, \nu_0(\varrho)).$$

Now, using the normality of $\mathscr{P}$ we have a constant $c > 0$, such that

$$\|\mathscr{F}(\varrho, x(\varrho))\| \le c, \ \ x \in I. \tag{1.7}$$

For convenience, the proof is divided into following steps:

Step I. In this step, we will show the continuity of the map $\mathscr{Q}$ on I. Consider a sequence $\{x_n\}$ in I such that $x_n \to x \in I$. The continuity of $\mathscr{F}$ implies that $\mathscr{F}(\varrho, x_n(\varrho)) \to \mathscr{F}(\varrho, x(\varrho))$ for $\varrho \in [s_i, t_{i+1}]$; $i = 0, 1, 2, \ldots, m$, similarly $\gamma_i(\varrho, x_n(\varrho)) \to \gamma_i(\varrho, x(\varrho))$ for $\varrho \in (t_i, s_i]$; $i = 1, 2, \ldots, m$ and from (1.7) we get that $\|\mathscr{F}(\varrho, x_n(\varrho)) - \mathscr{F}(\varrho, x(\varrho))\| \le 2c$. Combining these with Lebesgue dominated convergence theorem, we estimate the followings.
For $t \in [0, t_1]$,

$$\begin{aligned}\|\mathscr{Q}x_n(t) - \mathscr{Q}x(t)\| &\le \mathscr{M} \int_0^t \|\mathscr{F}(\varrho, x_n(\varrho)) - \mathscr{F}(\varrho, x(\varrho))\| d\varrho \\ &\to 0 \quad \text{as} \quad n \to \infty.\end{aligned}$$

For $t \in (t_i, s_i],\ i = 1, 2, \ldots, m$, we get

$$\begin{aligned}\|\mathscr{Q}x_n(t) - \mathscr{Q}x(t)\| &= \|\gamma_i(t, x_n(t)) - \gamma_i(t, x(t))\| \\ &\to 0 \quad \text{as} \quad n \to \infty.\end{aligned}$$

If $t \in (s_i, t_{i+1}];\ i = 1, 2, \ldots, m$,

$$\begin{aligned}\|\mathscr{Q}x_n(t) - \mathscr{Q}x(t)\| &\le \mathscr{M}\|\gamma_i(s_i, x_n(s_i)) - \gamma_i(s_i, x(s_i))\| \\ &\quad + \mathscr{M}\int_{s_i}^{t} \|\mathscr{F}(\varrho, x_n(\varrho)) - \mathscr{F}(\varrho, x(\varrho))\| d\varrho \\ &\to 0 \quad \text{as} \quad n \to \infty.\end{aligned}$$

Thus $\mathscr{Q}$ is continuous map on I.

Step II. The map $\mathscr{Q} : I \to I$ is monotone increasing. Let $x_1, x_2 \in I,\ x_1 \le x_2$. Using the positivity of $\mathscr{S}(t, s)$ and the hypotheses $(H1)$, $(H2)$, it is easy to see that $\mathscr{Q}x_1 \le \mathscr{Q}x_2$, which means $\mathscr{Q}$ is increasing operator. By Definition 1.4, we get $\omega_0 \le \mathscr{Q}\omega_0$. In the same way, we have $\mathscr{Q}\nu_0 \le \nu_0$. Let $u \in I$, so we have $\omega_0 \le \mathscr{Q}\omega_0 \le \mathscr{Q}u \le \mathscr{Q}\nu_0 \le \nu_0$, that means $\mathscr{Q}u \in I$. Therefore, $\mathscr{Q} : I \to I$ is monotone increasing.

Step III. We prove that $\mathscr{Q}$ has a fixed point. Consider the following sequences

$$\omega_n = \mathscr{Q}\omega_{n-1} \quad \text{and} \quad \nu_n = \mathscr{Q}\nu_{n-1}, \quad n \in N, \tag{1.8}$$

monotonicity of $\mathscr{Q}$ implies

$$\omega_0 \le \omega_1 \le \cdots \omega_n \le \cdots \le \nu_n \le \cdots \le \nu_1 \le \nu_0. \tag{1.9}$$

Let $S = \{\omega_n\}$ and $S_0 = \{\omega_{n-1}\}$. Then $S_0 = S \cup \{\omega_0\}$ and $\mu(S_0(t)) = \mu(S(t))$, $t \in J$. From (1.9), it is clear that S and S_0 are bounded. Observe that $\mu(\mathscr{S}(t, 0)(x_0)) = 0$, for $\{x_0\}$ is compact set and $\mathscr{S}(t, 0)$ is bounded. Suppose

$$\Phi(t) := \mu(S(t)), \ t \in J. \tag{1.10}$$

By the definition of measure of noncompactness, we have $\Phi(t) \ge 0$. For $t \in [0, t_1]$, from Lemma 1.1, Lemma 1.3, assumption $(H3)$, (1.6) and (1.8), we get

$$\begin{aligned}\Phi(t) &= \mu(\mathscr{Q}S_0(t)) \\ &= \mu\Big(\mathscr{S}(t, 0)x_0 + \int_0^t \mathscr{S}(t, \eta)\mathscr{F}(\eta, \omega_{n-1}(\eta))d\eta\Big) \\ &\le \mu(\mathscr{S}(t, 0)x_0) + 2\mathscr{M}\int_0^t \mu\Big(\mathscr{F}(\eta, \omega_{n-1}(\eta))\Big)d\eta\end{aligned}$$

$$
\begin{aligned}
&\le 2\mathscr{M}\mathscr{L}\int_0^t \mu(\{\omega_{n-1}(\eta)\})d\eta \\
&\le 2\mathscr{M}\mathscr{L}\int_0^t \mu(S_0(\eta))d\eta = 2\mathscr{M}\mathscr{L}\int_0^t \mu(S(\eta))d\eta \\
\Phi(t) &\le 2\mathscr{M}\mathscr{L}\int_0^t \Phi(\eta)d\eta.
\end{aligned}
$$

The above inequality combined with Lemma 1.4 imply that $\Phi(t) \equiv 0$ on $[0, t_1]$. For $t \in (t_i, s_i]$; $i = 1, 2, \ldots, m$, by using assumption $(H4)$

$$
\begin{aligned}
\Phi(t) &= \mu(\mathscr{Q}S_0(t)) \\
&= \mu(\{\gamma_i(t, \omega_{n-1}(t))\}) \\
&\le \mathscr{L}_i\mu(\omega_{n-1}(t)) = \mathscr{L}_i\mu(S(t)) = \mathscr{L}_i\Phi(t).
\end{aligned}
$$

Since $\mathscr{L}_i < 1$, therefore $\Phi(t) \equiv 0$ for all $t \in (t_i, s_i]$, $i = 1, 2, \ldots, m$. Observe that $S(s_i) = \{\omega_n(s_i)\}$ which is a monotone bounded sequence in X, therefore convergent for all $i = 1, 2, \ldots, m$. Thus $\mu(S(s_i)) = 0$ for all $i = 1, 2, \ldots, m$. Now, if $t \in (s_i, t_{i+1}]$; $i = 1, 2, \ldots, m$, from Lemma 1.1, Lemma 1.3 and assumptions $(H3)$, $(H4)$

$$
\begin{aligned}
\Phi(t) &= \mu(\mathscr{Q}S_0(t)) \\
&= \mu\left(\mathscr{S}(t, s_i)\gamma_i(s_i, \omega_{n-1}(s_i)) + \int_{s_i}^t \mathscr{S}(t, \eta)\mathscr{F}(\eta, \omega_{n-1}(\eta))d\eta\right) \\
&\le \mathscr{M}\mu(\{\gamma_i(s_i, \omega_{n-1}(s_i))\}) + 2\mathscr{M}\int_{s_i}^t \mu\left(\mathscr{F}(\eta, \omega_{n-1}(\eta))\right)d\eta \\
&\le \mathscr{M}\mathscr{L}_i\mu(\omega_{n-1}(s_i)) + 2\mathscr{M}\mathscr{L}\int_{s_i}^t \mu(\{\omega_{n-1}(\eta)\})d\eta \\
&= \mathscr{M}\mathscr{L}_i\mu(S(s_i)) + 2\mathscr{M}\mathscr{L}\int_{s_i}^t \Phi(\eta)d\eta \\
\Phi(t) &\le 2\mathscr{M}\mathscr{L}\int_{s_i}^t \Phi(\eta)d\eta.
\end{aligned}
$$

The above inequality combined with Lemma 1.4 imply that $\Phi(t) \equiv 0$ on $(s_i, t_{i+1}]$; $i = 1, 2, \ldots, m$. The above discussion concludes that $\mu(\{\omega_n(t)\}) = 0$ for every $t \in J$, this implies that the sequence $\{\omega_n(t)\}$ is precompact in X for each $t \in J$. Hence, we have a convergent subsequence of $\{\omega_n(t)\}$, combining this with (1.9) it is easy to observe that $\{\omega_n(t)\}$ itself is convergent in X. We denote $\lim_{n\to\infty} \omega_n(t) = \omega^*(t)$, $t \in J$. By (1.6) and (1.8), we obtain that

$$\omega_n(t) = \mathcal{Q}\omega_{n-1}(t) = \begin{cases} \mathcal{S}(t,0)x_0 + \int_0^t \mathcal{S}(t,\eta)\mathcal{F}(\eta,\omega_{n-1}(\eta))d\eta, & t \in [0,t_1], \\ \gamma_i(t,\omega_{n-1}(t)), \quad t \in \cup_{i=1}^m (t_i,s_i], \\ \mathcal{S}(t,s_i)\gamma_i(s_i,\omega_{n-1}(s_i)) + \int_{s_i}^t \mathcal{S}(t,\eta)\mathcal{F}(\eta,\omega_{n-1}(\eta))d\eta, \\ \quad t \in \cup_{i=1}^m (s_i,t_{i+1}]. \end{cases}$$

Taking $n \to \infty$ and applying Lebesgue dominated convergence theorem, we have

$$\omega^*(t) = \begin{cases} \mathcal{S}(t,0)x_0 + \int_0^t \mathcal{S}(t,\eta)\mathcal{F}(\eta,\omega^*(\eta))d\eta, & t \in [0,t_1], \\ \gamma_i(t,\omega^*(t)), \quad t \in \cup_{i=1}^m (t_i,s_i], \\ \mathcal{S}(t,s_i)\gamma_i(s_i,\omega^*(s_i)) + \int_{s_i}^t \mathcal{S}(t,\eta)\mathcal{F}(\eta,\omega^*(\eta))d\eta, \\ \quad t \in \cup_{i=1}^m (s_i,t_{i+1}]. \end{cases}$$

So, $\omega^* = \mathcal{Q}\omega^*$ and $\omega^* \in \mathcal{PC}(J,X)$. It proves that ω^* is a mild solution for the system (1.1). Similarly, there is $\nu^* \in \mathcal{PC}(J,X)$ satisfying $\nu^* = \mathcal{Q}\nu^*$. Now we show ω^*, ν^* are extremal mild solutions. Let $x \in I$ and $x = \mathcal{Q}x$, then $\omega_1 = \mathcal{Q}\omega_0 \le \mathcal{Q}x = x \le \mathcal{Q}\nu_0 = \nu_1$, by induction method, we obtain $\omega_n \le x \le \nu_n$. Hence $\omega_0 \le \omega^* \le x \le \nu^* \le \nu_0$ as $n \to \infty$, that means minimal and maximal mild solutions are ω^* and ν^*, respectively, for (1.1) in $[\omega_0,\nu_0]$.

□

To prove uniqueness, we need some more assumptions mentioned as below:

(H5) We have a constant $\mathcal{M}_1 > 0$ such that

$$\mathcal{F}(t,y_2) - \mathcal{F}(t,y_1) \le \mathcal{M}_1(y_2 - y_1),$$

where $y_1, y_2 \in X$ with $\omega_0(t) \le y_1 \le y_2 \le \nu_0(t)$, $t \in \cup_{i=0}^m [s_i,t_{i+1}]$.

(H6) There exist constants $\mathcal{N}_i > 0$ such that

$$\gamma_i(t,y_2) - \gamma_i(t,y_1) \le \mathcal{N}_i(y_2 - y_1),$$

for $t \in (t_i,s_i]$; $i = 1,2,\ldots,m$, and $y_1,y_2 \in X$ with $\omega_0(t) \le y_1 \le y_2 \le \nu_0(t)$.

Let us denote $\mathcal{N}^* = \max_{i=1,2,\ldots,m} \{\mathcal{N}_i\}$.

Theorem 1.4 *Suppose that the assumptions* $(H0)$, $(H1)$, $(H2)$, $(H5)$, $(H6)$, $(A1)$–$(A3)$ *hold and* $\omega_0, \nu_0 \in \mathcal{PC}(J,X)$ *with* $\omega_0 \le \nu_0$ *are lower and upper mild solutions, respectively, for the system* (1.1). *Then, there exists a unique mild solution for the system* (1.1) *in* $[\omega_0,\nu_0]$, *provided that* $\Lambda_1 := \mathcal{M}\mathcal{N}(\mathcal{N}^* + \mathcal{M}_1 b) < 1$.

Proof Let $\{y_n\} \subset [\omega_0(t),\nu_0(t)]$ be increasing monotone sequence. For $t \in \cup_{i=0}^m [s_i,t_{i+1}]$ and $n,m \in N$ $(n > m)$, the assumptions $(H1)$ and $(H5)$ imply

$$0 \leq \mathscr{F}(t, y_n) - \mathscr{F}(t, y_m) \leq \mathscr{M}_1(y_n - y_m).$$

Since the positive cone is normal, therefore

$$\|\mathscr{F}(t, y_n) - \mathscr{F}(t, y_m)\| \leq \mathscr{N}\mathscr{M}_1\|(y_n - y_m)\|.$$

Using Lemma 1.1, it yields

$$\mu(\{\mathscr{F}(t, y_n)\}) \leq \mathscr{N}\mathscr{M}_1\mu(\{y_n\}).$$

Hence the assumption $(H3)$ holds. Similarly, by $(H2)$, $(H6)$ and Lemma 1.1 we have

$$\mu(\{\gamma_i(t, y_n)\}) \leq \mathscr{N}\mathscr{N}_i\mu(\{y_n\}),$$

for each $t \in (t_i, s_i]$; $i = 1, 2, \ldots, m$. This means assumption $(H4)$ satisfies, and we can apply Theorem 1.3. Therefore, there exist minimal mild solution ω^* and maximal mild solutions ν^* for the system (1.1) in $[\omega_0, \nu_0]$.

For $t \in [0, t_1]$, by using (1.6), $(H5)$

$$\begin{aligned}
0 \leq \nu^*(t) - \omega^*(t) &= \mathscr{Q}\nu^*(t) - \mathscr{Q}\omega^*(t) \\
&= \int_0^t \mathscr{S}(t, \eta)\Big[\mathscr{F}(\eta, \nu^*(\eta)) - \mathscr{F}(\eta, \omega^*(\eta))\Big]d\eta \\
&\leq \mathscr{M}_1 \int_0^t \mathscr{S}(t, \eta)\Big(\nu^*(\eta) - \omega^*(\eta)\Big)d\eta.
\end{aligned}$$

Therefore, normality of cone implies

$$\begin{aligned}
\|\nu^* - \omega^*\| &\leq \mathscr{N}\left\|\mathscr{M}_1 \int_0^t \mathscr{S}(t, \eta)\Big(\nu^*(\eta) - \omega^*(\eta)\Big)d\eta\right\| \\
&\leq \mathscr{N}\mathscr{M}\mathscr{M}_1 b\|\nu^* - \omega^*\|. \qquad (1.11)
\end{aligned}$$

If $t \in (t_i, s_i]$, $i = 1, 2, \ldots, m$, assumption $(H6)$ and (1.6) yield

$$\begin{aligned}
0 \leq \nu^*(t) - \omega^*(t) &= \mathscr{Q}\nu^*(t) - \mathscr{Q}\omega^*(t) \\
&= \gamma_i(t, \nu^*(t)) - \gamma_i(t, \omega^*(t)) \\
&\leq \mathscr{N}_i(\nu^*(t) - \omega^*(t)).
\end{aligned}$$

Therefore, normality of cone implies

$$\begin{aligned}
\|\nu^* - \omega^*\| &\leq \mathscr{N}\mathscr{N}_i\|\nu^* - \omega^*\| \\
&\leq \mathscr{N}\mathscr{N}^*\|\nu^* - \omega^*\|. \qquad (1.12)
\end{aligned}$$

Similarly, if $t \in (s_i, t_{i+1}]$; $i = 1, 2, \ldots, m$, by (1.6), $(H5)$ and $(H6)$ we obtain

$$\begin{aligned}
0 \le \nu^*(t) - \omega^*(t) &= \mathscr{Q}\nu^*(t) - \mathscr{Q}\omega^*(t) \\
&= \mathscr{S}(t, s_i)\Big(\gamma_i(s_i, \nu^*(s_i)) - \gamma_i(s_i, \omega^*(s_i))\Big) \\
&\quad + \int_{s_i}^{t} \mathscr{S}(t, \eta)\Big(\mathscr{F}(\eta, \nu^*(\eta)) - \mathscr{F}(\eta, \omega^*(\eta))\Big)d\eta \\
&\le \mathscr{N}_i\Big[\mathscr{S}(t, s_i)(\nu^*(s_i) - \omega^*(s_i))\Big] \\
&\quad + \mathscr{M}_1 \int_{s_i}^{t} \mathscr{S}(t, \eta)(\nu^*(\eta) - \omega^*(\eta))d\eta.
\end{aligned}$$

The normality condition yields

$$\begin{aligned}
\|\nu^* - \omega^*\| &\le \mathscr{M}\mathscr{N}(\mathscr{N}_i + \mathscr{M}_1 b)\|\nu^* - \omega^*\| \\
&\le \mathscr{M}\mathscr{N}(\mathscr{N}^* + \mathscr{M}_1 b)\|\nu^* - \omega^*\|. \qquad (1.13)
\end{aligned}$$

The inequalities (1.11), (1.12) and (1.13) yield

$$\|\nu^* - \omega^*\| \le \Lambda_1 \|\nu^* - \omega^*\|.$$

Since $\Lambda_1 < 1$, so $\|\nu^* - \omega^*\| = 0$, that means $\nu^*(t) = \omega^*(t)$ on J. Thus, the uniqueness of the mild solution for (1.1) in $[\omega_0, \nu_0]$ is proved. □

1.4 Nonlocal Problem

The generalization of classical initial condition is known as nonlocal condition, it produces better results in the application of physical problems. The nonlocal Cauchy problem was first studied by Byszewski [4]. The nonlocal condition is used to describe the diffusion phenomenon of gas in a transparent tube by Deng [12]. In this section, we will discuss about extremal mild solutions for the non-autonomous non-instantaneous impulsive differential equations with nonlocal condition given as below:

$$\begin{aligned}
x'(t) + \mathscr{A}(t)x(t) &= \mathscr{F}(t, x(t)), \quad t \in \cup_{i=0}^{m}(s_i, t_{i+1}], \\
x(t) &= \gamma_i(t, x(t)), \quad t \in \cup_{i=1}^{m}(t_i, s_i], \\
x(0) &= x_0 + \mathscr{G}(x). \qquad (1.14)
\end{aligned}$$

Definition 1.6 A function $x \in \mathscr{PC}(J, X)$ satisfying the following integral equation

$$x(t) = \begin{cases} \mathscr{S}(t,0)(x_0 + \mathscr{G}(x)) + \int_0^t \mathscr{S}(t,\eta)\mathscr{F}(\eta, x(\eta))d\eta, & 0 \le t \le t_1, \\ \gamma_i(t, x(t)), \qquad t \in \cup_{i=1}^m (t_i, s_i], \\ \mathscr{S}(t, s_i)\gamma_i(s_i, x(s_i)) + \int_{s_i}^t \mathscr{S}(t,\eta)\mathscr{F}(\eta, x(\eta))d\eta, & t \in \cup_{i=1}^m (s_i, t_{i+1}], \end{cases}$$

is named as mild solution of the problem (1.14).

Definition 1.7 $\omega_0 \in \mathscr{PC}(J, X)$ satisfying the following

$$\omega_0(t) \le \begin{cases} \mathscr{S}(t,0)(x_0 + \mathscr{G}(\omega_0)) + \int_0^t \mathscr{S}(t,\eta)\mathscr{F}(\eta, \omega_0(\eta))d\eta, & 0 \le t \le t_1, \\ \gamma_i(t, \omega_0(t)), \qquad t \in \cup_{i=1}^m (t_i, s_i], \\ \mathscr{S}(t, s_i)\gamma_i(s_i, \omega_0(s_i)) + \int_{s_i}^t \mathscr{S}(t,\eta)\mathscr{F}(\eta, \omega_0(\eta))d\eta, & t \in \cup_{i=1}^m (s_i, t_{i+1}], \end{cases} \tag{1.15}$$

is named as lower mild solution for the system (1.14). If the inequalities of (1.15) are opposite, solution is named as upper mild solution.

To prove the existence and uniqueness of extremal mild solutions of nonlocal problem (1.14), we need following conditions on nonlocal function:

(H7) $\mathscr{G}$ is X-valued continuous increasing compact function defined on $\mathscr{PC}(J, X)$.

(H8) $\mathscr{G}$ satisfies

$$\mathscr{G}(y) - \mathscr{G}(x) \le \mathscr{L}_g(y - x), \text{ for } x, y \in I \text{ with } x \le y,$$

for some constant $\mathscr{L}_g > 0$.

Let us define the operator $\mathscr{Q}$ on $\mathscr{PC}(J, X)$ as following:

$$\mathscr{Q}x(t) = \begin{cases} \mathscr{S}(t,0)(x_0 + \mathscr{G}(x)) + \int_0^t \mathscr{S}(t,\eta)\mathscr{F}(\eta, x(\eta))d\eta, & 0 \le t \le t_1, \\ \gamma_i(t, x(t)), \qquad t \in \cup_{i=1}^m (t_i, s_i], \\ \mathscr{S}(t, s_i)\gamma_i(s_i, x(s_i)) + \int_{s_i}^t \mathscr{S}(t,\eta)\mathscr{F}(\eta, x(\eta))d\eta, & t \in \cup_{i=1}^m (s_i, t_{i+1}]. \end{cases} \tag{1.16}$$

Theorem 1.5 *If the assumptions* $(A1)$–$(A3)$, $(H0)$–$(H4)$ *and* $(H7)$ *are satisfied, and* $\omega_0, \nu_0 \in \mathscr{PC}(J, X)$ *with* $\omega_0 \le \nu_0$ *are lower and upper mild solutions, respectively, for the system* (1.14). *Then, there exist extremal mild solutions for the problem* (1.14) *in the interval* $[\omega_0, \nu_0]$, *provided that* $\max_{i=1,2,\ldots,m} \{\mathscr{L}_i\} < 1$.

Proof We can check easily that $\mathscr{Q} : I \to I$ is continuous and increasing. Consider the following sequences

$$\omega_n = \mathscr{Q}\omega_{n-1} \quad \text{and} \quad \nu_n = \mathscr{Q}\nu_{n-1}, \quad n \in N, \tag{1.17}$$

monotonicity of $\mathscr{Q}$ implies

$$\omega_0 \le \omega_1 \le \cdots \omega_n \le \cdots \le \nu_n \le \cdots \le \nu_1 \le \nu_0. \tag{1.18}$$

Let $S = \{\omega_n\}$ and $S_0 = \{\omega_{n-1}\}$. Then $S_0 = S \cup \{\omega_0\}$ and $\mu(S_0(t)) = \mu(S(t))$, $t \in J$. From (1.18), it is clear that S and S_0 are bounded. Observe that $\mu(\mathscr{S}(t,0)(x_0)) = 0$, $\mu(\mathscr{S}(t,0)\mathscr{G}(\omega_{n-1})) = 0$ for $\{x_0\}$ is compact set, $\mathscr{G}$ is compact map and $\mathscr{S}(t,0)$ is bounded. Suppose

$$\Phi(t) := \mu(S(t)), \ t \in J. \tag{1.19}$$

For $t \in [0, t_1]$, from Lemma 1.1, Lemma 1.3, assumption $(H3)$, (1.16) and (1.17), we get

$$\begin{aligned}
\Phi(t) &= \mu(\mathscr{Q}S_0(t)) \\
&= \mu\left(\mathscr{S}(t,0)(x_0 + \mathscr{G}(\omega_{n-1})) + \int_0^t \mathscr{S}(t,\eta)\mathscr{F}(\eta, \omega_{n-1}(\eta))d\eta\right) \\
&\leq \mu(\mathscr{S}(t,0)x_0) + \mu(\mathscr{S}(t,0)\mathscr{G}(\omega_{n-1})) + 2\mathscr{M}\int_0^t \mu\left(\mathscr{F}(\eta, \omega_{n-1}(\eta))\right)d\eta \\
&\leq 2\mathscr{M}\mathscr{L}\int_0^t \mu(\{\omega_{n-1}(\eta)\})d\eta \\
&\leq 2\mathscr{M}\mathscr{L}\int_0^t \mu(S_0(\eta))d\eta = 2\mathscr{M}\mathscr{L}\int_0^t \mu(S(\eta))d\eta \\
\Phi(t) &\leq 2\mathscr{M}\mathscr{L}\int_0^t \Phi(\eta)d\eta.
\end{aligned}$$

Combining the above inequality with Lemma 1.4 imply that $\Phi(t) \equiv 0$ on $[0, t_1]$. Proceeding in the same way as the proof of Theorem 1.3, we obtain

$$\begin{aligned}
\Phi(t) &\equiv 0, \quad t \in \cup_{i=1}^m (t_i, s_i]. \\
\Phi(t) &\equiv 0, \quad t \in \cup_{i=1}^m (s_i, t_{i+1}](s_i, t_{i+1}].
\end{aligned}$$

By the above discussion, we conclude that $\mu(\{\omega_n(t)\}) = 0$ for every $t \in J$, this implies that the sequence $\{\omega_n(t)\}$ is precompact in X for each $t \in J$. Therefore, it has a convergent subsequence, combining this with (1.18) yield that $\{\omega_n(t)\}$ is convergent in X. Denote $\lim_{n\to\infty} \omega_n(t) = \omega^*(t)$, $t \in J$. By (1.16) and (1.17), we obtain that

$$\omega_n(t) = \mathscr{Q}\omega_{n-1}(t) = \begin{cases} \mathscr{S}(t,0)(x_0 + \mathscr{G}\omega_{n-1}) + \int_0^t \mathscr{S}(t,\eta)\mathscr{F}(\eta, \omega_{n-1}(\eta))d\eta, \\ \quad 0 \leq t \leq t_1, \\ \gamma_i(t, \omega_{n-1}(t)), \quad\quad t \in \cup_{i=1}^m (t_i, s_i], \\ \mathscr{S}(t, s_i)\gamma_i(s_i, \omega_{n-1}(s_i)) + \int_{s_i}^t \mathscr{S}(t,\eta)\mathscr{F}(\eta, \omega_{n-1}(\eta))d\eta, \\ \quad t \in \cup_{i=1}^m (s_i, t_{i+1}]. \end{cases}$$

Taking $n \to \infty$ and applying Lebesgue dominated convergence theorem, we have

$$\omega^*(t) = \begin{cases} \mathscr{S}(t,0)(x_0 + \mathscr{G}\omega^*) + \int_0^t \mathscr{S}(t,\eta)\mathscr{F}(\eta,\omega^*(\eta))d\eta, & 0 \le t \le t_1, \\ \gamma_i(t,\omega^*(t)), \quad t \in \cup_{i=1}^m (t_i, s_i], \\ \mathscr{S}(t,s_i)\gamma_i(s_i,\omega^*(s_i)) + \int_{s_i}^t \mathscr{S}(t,\eta)\mathscr{F}(\eta,\omega^*(\eta))d\eta, \\ \quad t \in \cup_{i=1}^m (s_i, t_{i+1}]. \end{cases}$$

So, $\omega^* = \mathscr{Q}\omega^*$ and $\omega^* \in \mathscr{PC}(J,X)$, which imply that ω^* is a mild solution of (1.14). Similarly, we have $\nu^* \in \mathscr{PC}(J,X)$ and $\nu^* = \mathscr{Q}\nu^*$. Now we show ω^*, ν^* are extremal mild solutions. Let $x \in I$ and $x = \mathscr{Q}x$, then $\omega_1 = \mathscr{Q}\omega_0 \le \mathscr{Q}x = x \le \mathscr{Q}\nu_0 = \nu_1$, applying the method of induction $\omega_n \le x \le \nu_n$, hence $\omega_0 \le \omega^* \le x \le \nu^* \le \nu_0$. It means that minimal and maximal mild solutions are ω^* and ν^* for (1.14) in $[\omega_0, \nu_0]$. □

Theorem 1.6 *If the assumptions* $(H0)$–$(H2)$, $(H5)$–$(H8)$, $(A1)$–$(A3)$ *hold and* $\omega_0, \nu_0 \in \mathscr{PC}(J,X)$ *with* $\omega_0 \le \nu_0$ *are lower and upper mild solutions, respectively, to the system* (1.14). *Then, the system* (1.14) *has a unique mild solution in* $[\omega_0, \nu_0]$, *provided that*

$$\Lambda_2 := \max\left\{\mathscr{M}\mathscr{N}(\mathscr{L}_g + \mathscr{M}_1 b),\ \mathscr{M}\mathscr{N}(\mathscr{N}^* + \mathscr{M}_1 b)\right\} < 1.$$

Proof We can easily see that $(H3)$ and $(H4)$ hold, as it is done in the proof of Theorem 1.4. Now, applying Theorem 1.5, we get that the system (1.14) has minimal mild solution ω^* and maximal mild solutions ν^* in $[\omega_0, \nu_0]$.

For $t \in [0, t_1]$, by using (1.16), $(H5)$ and $(H8)$, we estimate

$$\begin{aligned} 0 &\le \nu^*(t) - \omega^*(t) = \mathscr{Q}\nu^*(t) - \mathscr{Q}\omega^*(t) \\ &= \mathscr{S}(t,0)(\mathscr{G}(\nu^*) - \mathscr{G}(\omega^*)) + \int_0^t \mathscr{S}(t,\eta)\Big[\mathscr{F}(\eta,\nu^*(\eta)) - \mathscr{F}(\eta,\omega^*(\eta))\Big]d\eta \\ &\le \mathscr{L}_g\mathscr{S}(t,0)(\nu^* - \omega^*) + \mathscr{M}_1 \int_0^t \mathscr{S}(t,\eta)\Big(\nu^*(\eta) - \omega^*(\eta)\Big)d\eta. \end{aligned}$$

Therefore, normality of cone implies

$$\|\nu^* - \omega^*\| \le \mathscr{N}\mathscr{M}\Big[\mathscr{L}_g + \mathscr{M}_1 b\Big]\|\nu^* - \omega^*\|. \tag{1.20}$$

Also, we have

$$\|\nu^* - \omega^*\| \le \mathscr{N}\mathscr{N}^*\|\nu^* - \omega^*\|, \quad t \in \cup_{i=1}^m (t_i, s_i], \tag{1.21}$$

and

$$\|\nu^* - \omega^*\| \le \mathscr{M}\mathscr{N}(\mathscr{N}^* + \mathscr{M}_1 b)\|\nu^* - \omega^*\|, \quad t \in \cup_{i=1}^m (s_i, t_{i+1}]. \tag{1.22}$$

The inequalities (1.20), (1.21) and (1.22) yield

$$\|\nu^* - \omega^*\| \leq \Lambda_2 \|\nu^* - \omega^*\|.$$

Since $\Lambda_2 < 1$, so $\|\nu^* - \omega^*\| = 0$, which means $\nu^*(t) = \omega^*(t)$, on J. Thus, the uniqueness of mild solution for (1.14) in $[\omega_0, \nu_0]$ is proved. □

1.5 Integro-Differential Equations

This section is concerned with non-autonomous integro-differential equations with non-instantaneous impulsive conditions:

$$\begin{aligned} x'(t) + \mathscr{A}(t)x(t) &= \mathscr{F}\left(t, x(t), \int_0^t k(t,s)x(s)ds\right), \quad t \in \cup_{i=0}^m (s_i, t_{i+1}], \\ x(t) &= \gamma_i(t, x(t)), \quad t \in \cup_{i=1}^m (t_i, s_i], \\ x(0) &= x_0, \end{aligned} \tag{1.23}$$

$k \in \mathscr{C}(D, R^+)$ with $D := \{(\tau, s) : 0 \leq s \leq \tau \leq b\}$. For our convenience we denote $\mathscr{K}x(t) := \int_0^t k(t,s)x(s)ds$, and $K^* := \sup_{(t,s)\in D} k(t,s)$.

Definition 1.8 A function $x \in \mathscr{PC}(J, X)$ is said to be a mild solution of the problem (1.23) if it satisfies

$$x(t) = \begin{cases} \mathscr{S}(t,0)x_0 + \int_0^t \mathscr{S}(t,\eta)\mathscr{F}(\eta, x(\eta), \mathscr{K}x(\eta))d\eta, \\ \qquad 0 \leq t \leq t_1, \\ \gamma_i(t, x(t)), \qquad t \in \cup_{i=1}^m (t_i, s_i], \\ \mathscr{S}(t, s_i)\gamma_i(s_i, x(s_i)) + \int_{s_i}^t \mathscr{S}(t,\eta)\mathscr{F}(\eta, x(\eta), \mathscr{K}x(\eta))d\eta, \\ \qquad t \in \cup_{i=1}^m (s_i, t_{i+1}]. \end{cases}$$

Definition 1.9 $\omega_0 \in \mathscr{PC}(J, X)$ satisfying

$$\omega_0(t) \leq \begin{cases} \mathscr{S}(t,0)x_0 + \int_0^t \mathscr{S}(t,\eta)\mathscr{F}(\eta, \omega_0(\eta), \mathscr{K}\omega_0(\eta))d\eta, \\ \qquad t \in [0, t_1], \\ \gamma_i(t, \omega_0(t)), \qquad t \in \cup_{i=1}^m (t_i, s_i], \\ \mathscr{S}(t, s_i)\gamma_i(s_i, \omega_0(s_i)) + \int_{s_i}^t \mathscr{S}(t,\eta)\mathscr{F}(\eta, \omega_0(\eta), \mathscr{K}\omega_0(\eta))d\eta, \\ \qquad t \in \cup_{i=1}^m (s_i, t_{i+1}], \end{cases} \tag{1.24}$$

is named as lower mild solution for the system (1.23). If the inequalities of (1.24) are opposite, solution is named as upper mild solution.

Let $\mathscr{Q} : \mathscr{PC}(J, X) \to \mathscr{PC}(J, X)$ as following:

$$\mathscr{Q}x(t) = \begin{cases} \mathscr{S}(t,0)x_0 + \int_0^t \mathscr{S}(t,\eta)\mathscr{F}(\eta, x(\eta), \mathscr{K}x(\eta))d\eta, \\ \quad t \in [0, t_1], \\ \gamma_i(t, x(t)), \qquad t \in \cup_{i=1}^m (t_i, s_i], \\ \mathscr{S}(t, s_i)\gamma_i(s_i, x(s_i)) + \int_{s_i}^t \mathscr{S}(t,\eta)\mathscr{F}(\eta, x(\eta), \mathscr{K}x(\eta))d\eta, \\ \quad t \in \cup_{i=1}^m (s_i, t_{i+1}]. \end{cases} \tag{1.25}$$

Now, we state some more assumptions required to prove the existence of extremal mild solutions for the system (1.23):

(B1) The function $\mathscr{F}$ is continuous on $\cup_{i=0}^m [s_i, t_{i+1}] \times X \times X$. For $y_1, y_2 \in X$ with $\omega_0(t) \le y_1 \le y_2 \le \nu_0(t)$ and $\mathscr{K}\omega_0(t) \le x_1 \le x_2 \le \mathscr{K}\nu_0(t)$, we assumed

$$\mathscr{F}(t, y_1, x_1) \le \mathscr{F}(t, y_2, x_2).$$

(B2) For all $t \in \cup_{i=0}^m [s_i, t_{i+1}]$ and monotone increasing or decreasing sequences $\{y_n\} \subset [\omega_0(t), \nu_0(t)]$, $\{x_n\} \subset [\mathscr{K}\omega_0(t), \mathscr{K}\nu_0(t)]$, we have

$$\mu(\{\mathscr{F}(t, y_n, x_n)\}) \le \mathscr{L}(\mu(\{y_n\}) + \mu(\{x_n\})),$$

for some constant $\mathscr{L} > 0$.

Theorem 1.7 *If the assumptions* $(A1)$–$(A3)$ *and* $(B1)$, $(B2)$, $(H0)$, $(H2)$, $(H4)$ *are satisfied, and* $\omega_0, \nu_0 \in \mathscr{PC}(J, X)$ *with* $\omega_0 \le \nu_0$ *are lower and upper mild solutions, respectively, for* (1.23)*. Then, extremal mild solutions exist for the system* (1.23) *in the interval* $[\omega_0, \nu_0]$*, provided that* $\max_{i=1,2,\ldots,m} \{\mathscr{L}_i\} < 1$.

Proof Easily, we can check that $\mathscr{Q} : I \to I$ is continuous and increasing. Consider the following sequences

$$\omega_n = \mathscr{Q}\omega_{n-1} \quad \text{and} \quad \nu_n = \mathscr{Q}\nu_{n-1}, \quad n \in N, \tag{1.26}$$

monotonicity of $\mathscr{Q}$ implies

$$\omega_0 \le \omega_1 \le \cdots \omega_n \le \cdots \le \nu_n \le \cdots \le \nu_1 \le \nu_0. \tag{1.27}$$

Let $S = \{\omega_n\}$ and $S_0 = \{\omega_{n-1}\}$. Then $S_0 = S \cup \{\omega_0\}$ and $\mu(S_0(t)) = \mu(S(t))$, $t \in J$. From (1.27), it is clear that S and S_0 are bounded. Observe that $\mu(\mathscr{S}(t,0)(x_0)) = 0$, for $\{x_0\}$ is compact set and $\mathscr{S}(t,0)$ is bounded. Suppose

$$\Phi(t) := \mu(S(t)), \ t \in J. \tag{1.28}$$

With the help of Lemma 1.3, notice that

$$\mu\Big(\{\mathscr{K}\omega_{n-1}(\eta)\}\Big) = \mu\Big(\int_0^\eta k(\eta, s)\omega_{n-1}(s)ds\Big)$$

$$\begin{aligned}
&\leq K^*\mu\left(\int_0^{\eta}\omega_{n-1}(s)ds\right)\\
&\leq 2K^*\int_0^{\eta}\mu(\omega_{n-1}(s))ds\\
&\leq 2K^*\int_0^{\eta}\mu(S_0(s))ds = 2K^*\int_0^{\eta}\mu(S(s))ds\\
&\leq 2K^*\int_0^{\eta}\Phi(s)ds,
\end{aligned}$$

therefore

$$\int_0^t \mu\left(\{\mathscr{K}\omega_{n-1}(\eta)\}\right)d\eta \leq 2bK^*\int_0^t \Phi(\eta)d\eta. \tag{1.29}$$

For $t \in [0, t_1]$, from Lemma 1.1, Lemma 1.3, assumption $(B2)$, (1.25), (1.26) and (1.29), we get

$$\begin{aligned}
\Phi(t) &= \mu(\mathscr{Q}S_0(t))\\
&= \mu\left(\mathscr{S}(t,0)x_0 + \int_0^t \mathscr{S}(t,\eta)\mathscr{F}(\eta,\omega_{n-1}(\eta),\mathscr{K}\omega_{n-1}(\eta))d\eta\right)\\
&\leq \mu(\mathscr{S}(t,0)x_0) + 2\mathscr{M}\int_0^t \mu\left(\mathscr{F}(\eta,\omega_{n-1}(\eta),\mathscr{K}\omega_{n-1}(\eta))\right)d\eta\\
&\leq 2\mathscr{M}\mathscr{L}\int_0^t\left[\mu(\{\omega_{n-1}(\eta)\}) + \mu(\{\mathscr{K}\omega_{n-1}(\eta)\})\right]d\eta\\
&\leq 2\mathscr{M}\mathscr{L}\left[\int_0^t \mu(S(\eta))d\eta + 2bK^*\int_0^t\Phi(\eta)d\eta\right]\\
&\leq 2\mathscr{M}\mathscr{L}\left[\int_0^t \Phi(\eta)d\eta + 2bK^*\int_0^t\Phi(\eta)d\eta\right]\\
&\leq 2\mathscr{M}\mathscr{L}(1+2bK^*)\int_0^t\Phi(\eta)d\eta.
\end{aligned}$$

Combining the above inequality with Lemma 1.4, we conclude that $\Phi(t) \equiv 0$ for all $t \in [0, t_1]$. Also $\Phi(t) \equiv 0$ for all $t \in (t_i, s_i]$, $i = 1, 2, \ldots, m$, the proof is same as it is done in Theorem 1.3. Observe that $S(s_i) = \{\omega_n(s_i)\}$ which is a monotone bounded sequence in X, therefore convergent for all $i = 1, 2, \ldots, m$. Thus $\mu(S(s_i)) = 0$ for all $i = 1, 2, \ldots, m$. Now, for $t \in (s_i, t_{i=1}]$; $i = 1, 2, \ldots, m$, from Lemma 1.1, Lemma 1.3, $(B2)$, $(H4)$ and (1.29), we obtain

$$
\begin{aligned}
\Phi(t) &= \mu(\mathcal{Q}S_0(t)) \\
&= \mu\left(\mathcal{S}(t, s_i)\gamma_i(s_i, \omega_{n-1}(s_i)) + \int_{s_i}^{t} \mathcal{S}(t, \eta)\mathcal{F}(\eta, \omega_{n-1}(\eta), \mathcal{K}\omega_{n-1}(\eta))d\eta\right) \\
&\leq \mathcal{M}\mu(\{\gamma_i(s_i, \omega_{n-1}(s_i))\}) + 2\mathcal{M}\int_{s_i}^{t} \mu\left(\mathcal{F}(\eta, \omega_{n-1}(\eta), \mathcal{K}\omega_{n-1}(\eta))\right)d\eta \\
&\leq \mathcal{M}\mathcal{L}_i\mu(\omega_{n-1}(s_i)) + 2\mathcal{M}\mathcal{L}\int_{s_i}^{t}\left[\mu(\{\omega_{n-1}(\eta)\}) + \mu(\{\mathcal{K}\omega_{n-1}(\eta)\})\right]d\eta \\
&= \mathcal{M}\mathcal{L}_i\mu(S(s_i)) + 2\mathcal{M}\mathcal{L}\left[\int_{s_i}^{t}\mu(S(\eta))d\eta + 2bK^*\int_{s_i}^{t}\Phi(\eta)d\eta\right] \\
\Phi(t) &\leq 2\mathcal{M}\mathcal{L}\left[\int_{s_i}^{t}\Phi(\eta)d\eta + 2bK^*\int_{s_i}^{t}\Phi(\eta)d\eta\right] \\
&\leq 2\mathcal{M}\mathcal{L}(1 + 2bK^*)\int_{s_i}^{t}\Phi(\eta)d\eta.
\end{aligned}
$$

Combining the above inequality with Lemma 1.4, we conclude that $\Phi(t) \equiv 0$ for all $t \in (s_i, t_{i=1}]$, $i = 1, 2, \ldots, m$. The above discussion concludes that $\mu(\{\omega_n(t)\}) = 0$ for every $t \in J$, this implies that the sequence $\{\omega_n(t)\}$ is precompact in X for all $t \in J$. So, $\{\omega_n(t)\}$ has a convergent subsequence, combining this with (1.27) yields that $\{\omega_n(t)\}$ is convergent in X. Denote $\lim_{n\to\infty} \omega_n(t) = \omega^*(t),\ t \in J$. By (1.25) and (1.26), we obtain that

$$
\omega_n(t) = \mathcal{Q}\omega_{n-1}(t) = \begin{cases} \mathcal{S}(t, 0)x_0 + \int_0^t \mathcal{S}(t, \eta)\mathcal{F}(\eta, \omega_{n-1}(\eta), \mathcal{K}\omega_{n-1}(\eta))d\eta, \\ \quad 0 \leq t \leq t_1, \\ \gamma_i(t, \omega_{n-1}(t)), \qquad t \in \cup_{i=1}^m (t_i, s_i], \\ \mathcal{S}(t, s_i)\gamma_i(s_i, \omega_{n-1}(s_i)) + \\ \int_{s_i}^t \mathcal{S}(t, \eta)\mathcal{F}(\eta, \omega_{n-1}(\eta), \mathcal{K}\omega_{n-1}(\eta))d\eta,\ t \in \cup_{i=1}^m (s_i, t_{i+1}]. \end{cases}
$$

Taking $n \to \infty$ and applying Lebesgue dominated convergence theorem

$$
\omega^*(t) = \begin{cases} \mathcal{S}(t, 0)x_0 + \int_0^t \mathcal{S}(t, \eta)\mathcal{F}(\eta, \omega^*(\eta), \mathcal{K}\omega^*(\eta))d\eta, \\ \quad 0 \leq t \leq t_1, \\ \gamma_i(t, \omega^*(t)), \qquad t \in \cup_{i=1}^m (t_i, s_i], \\ \mathcal{S}(t, s_i)\gamma_i(s_i, \omega^*(s_i)) + \int_{s_i}^t \mathcal{S}(t, \eta)\mathcal{F}(\eta, \omega^*(\eta), \mathcal{K}\omega^*(\eta))d\eta, \\ \quad t \in \cup_{i=1}^m (s_i, t_{i+1}]. \end{cases}
$$

So, $\omega^* \in \mathcal{PC}(J, X)$ and $\omega^* = \mathcal{Q}\omega^*$, which yield that a mild solution for (1.23) is ω^*. Similarly, we obtain $\nu^* \in \mathcal{PC}(J, X)$ and $\nu^* = \mathcal{Q}\nu^*$. Now we show ω^*, ν^* are extremal mild solutions. Let $x \in I$ and $x = \mathcal{Q}x$, then $\omega_1 = \mathcal{Q}\omega_0 \leq \mathcal{Q}x = x \leq \mathcal{Q}\nu_0 = \nu_1$, by the method of induction $\omega_n \leq x \leq \nu_n$, hence $\omega_0 \leq \omega^* \leq x \leq \nu^* \leq \nu_0$, which means minimal and maximal mild solutions for (1.23) are ω^* and ν^* in $[\omega_0, \nu_0]$. □

To prove the uniqueness, we need one more assumption mentioned as following:

(B3) We have a constant $\mathscr{M}_1 > 0$ such that, for $t \in \cup_{i=0}^{m}[s_i, t_{i+1}]$

$$\mathscr{F}(t, y_2, x_2) - \mathscr{F}(t, y_1, x_1) \le \mathscr{M}_1[(y_2 - y_1) + (x_2 - x_1)],$$

where $y_1, y_2 \in X$ with $\omega_0(t) \le y_1 \le y_2 \le \nu_0(t)$ and $\mathscr{K}\omega_0(t) \le x_1 \le x_2 \le \mathscr{K}\nu_0(t)$.

Theorem 1.8 *If the assumptions* $(H0), (B1), (H2), (B3), (H6), (A1)$–$(A3)$ *hold and* $\omega_0, \nu_0 \in \mathscr{PC}(J, X)$ *with* $\omega_0 \le \nu_0$ *are lower and upper mild solutions, respectively, for* (1.23). *Then, the system* (1.23) *has a unique mild solution in* $[\omega_0, \nu_0]$, *provided that* $\Lambda_3 := \mathscr{M}\mathscr{N}\left[\mathscr{N}^* + \mathscr{M}_1 b(1 + bK^*)\right] < 1$.

Proof Let $\{y_n\} \subset [\omega_0(t), \nu_0(t)]$ and $\{x_n\} \subset [\mathscr{K}\omega_0(t), \mathscr{K}\nu_0(t)]$ be increasing monotone sequences. For $t \in \cup_{i=0}^{m}[s_i, t_{i+1}]$ and $n, m \in N$ $(n > m)$, the assumptions $(B1)$ and $(B3)$ yield

$$0 \le \mathscr{F}(t, y_n, x_n) - \mathscr{F}(t, y_m, x_m) \le \mathscr{M}_1[(y_n - y_m) + (x_n - x_m)].$$

Since the positive cone is normal, therefore

$$\|\mathscr{F}(t, y_n, x_n) - \mathscr{F}(t, y_m, x_m)\| \le \mathscr{N}\mathscr{M}_1\|(y_n - y_m) + (x_n - x_m)\|. \tag{1.30}$$

So, Lemma 1.1 implies

$$\mu(\{\mathscr{F}(t, y_n, x_n)\}) \le \mathscr{N}\mathscr{M}_1(\mu(\{y_n\}) + \mu(\{x_n\})).$$

Hence the assumption $(B2)$ holds. Similarly, by $(H2)$, $(H6)$ and Lemma 1.1 we have

$$\mu(\{\gamma_i(t, y_n)\}) \le \mathscr{N}\mathscr{N}_i\mu(\{y_n\}),$$

for each $t \in (t_i, s_i]$; $i = 1, 2, \ldots, m$. This means assumption $(H4)$ holds. Therefore, by Theorem 1.7 there exist minimal and maximal mild solutions ω^* ν^* in $[\omega_0, \nu_0]$ for the problem (1.23).

For $t \in [0, t_1]$, by using (1.25) and $(B3)$

$$\begin{aligned} 0 &\le \nu^*(t) - \omega^*(t) = \mathscr{Q}\nu^*(t) - \mathscr{Q}\omega^*(t) \\ &= \int_0^t \mathscr{S}(t, \eta)[\mathscr{F}(\eta, \nu^*(\eta), \mathscr{K}\nu^*(\eta)) - \mathscr{F}(\eta, \omega^*(\eta), \mathscr{K}\omega^*(\eta))]d\eta \\ &\le \mathscr{M}_1 \int_0^t \mathscr{S}(t, \eta)\left[(\nu^*(\eta) - \omega^*(\eta)) + (\mathscr{K}\nu^*(\eta) - \mathscr{K}\omega^*(\eta))\right]d\eta. \end{aligned}$$

Since the positive cone is normal, therefore

$$\begin{aligned}\|\nu^* - \omega^*\| &\le \mathscr{N}\mathscr{M}\mathscr{M}_1 b\Big(\|\nu^* - \omega^*\| + \|\mathscr{K}\nu^* - \mathscr{K}\omega^*\|\Big)\\ &\le \mathscr{N}\mathscr{M}\mathscr{M}_1 b(1 + bK^*)\|\nu^* - \omega^*\|. \end{aligned} \tag{1.31}$$

Also, we get

$$\|\nu^* - \omega^*\| \le \mathscr{N}\mathscr{N}^*\|\nu^* - \omega^*\|, \quad t \in \cup_{i=1}^m (t_i, s_i]. \tag{1.32}$$

If $t \in (s_i, t_{i+1}]$; $i = 1, 2, \ldots, m$, by (1.25), $(B3)$ and $(H6)$, we obtain

$$\begin{aligned}0 &\le \nu^*(t) - \omega^*(t) = \mathscr{Q}\nu^*(t) - \mathscr{Q}\omega^*(t)\\ &= \mathscr{S}(t, s_i)\Big(\gamma_i(s_i, \nu^*(s_i)) - \gamma_i(s_i, \omega^*(s_i))\Big)\\ &\quad + \int_{s_i}^t \mathscr{S}(t, \eta)\Big(\mathscr{F}(\eta, \nu^*(\eta), \mathscr{K}\nu^*(\eta)) - \mathscr{F}(\eta, \omega^*(\eta), \mathscr{K}\omega^*(\eta))\Big)d\eta\\ &\le \mathscr{N}_i\Big[\mathscr{S}(t, s_i)(\nu^*(s_i) - \omega^*(s_i))\Big]\\ &\quad + \mathscr{M}_1 \int_{s_i}^t \mathscr{S}(t, s_i)\Big[(\nu^*(\eta) - \omega^*(\eta)) + (\mathscr{K}\nu^*(\eta) - \mathscr{K}\omega^*(\eta))\Big]d\eta.\end{aligned}$$

Using the normality condition

$$\|\nu^* - \omega^*\| \le \mathscr{M}\mathscr{N}\Big[\mathscr{N}^* + \mathscr{M}_1 b(1 + bK^*)\Big]\|\nu^* - \omega^*\|. \tag{1.33}$$

The inequalities (1.31), (1.32) and (1.33) yield

$$\|\nu^* - \omega^*\| \le \Lambda_3\|\nu^* - \omega^*\|.$$

Since $\Lambda_3 < 1$, so $\|\nu^* - \omega^*\| = 0$, that means $\nu^*(t) = \omega^*(t)$ on J. Thus, the uniqueness of mild solution for (1.23) is proved in $[\omega_0, \nu_0]$. □

1.6 Example

We consider the following partial differential equation to illustrate our results:

$$\begin{cases} x'(t,y)+a(t,y)\frac{\partial^2}{\partial y^2}x(t,y)=\frac{e^{-t}}{49+e^t}x(t,y)+\int_0^t\frac{1}{50}e^{-s}x(s,y)ds, \\ \quad y\in[0,\pi],\ t\in(0,\frac{1}{3}]\cup(\frac{1}{2},1], \\ x(t,0)=0,\ x(t,\pi)=0 \qquad t\in J=[0,1], \\ x(t,y)=L_1e^{-(t-\frac{1}{3})}x(t,y), \qquad y\in(0,\pi), t\in(\frac{1}{3},\frac{1}{2}], \\ x(0,y)=\frac{|x(t,y)|}{7+|x(t,y)|}+x_0(y), \qquad y\in[0,\pi], \end{cases} \tag{1.34}$$

with $X=\mathscr{L}^2([0,\pi]\times[0,1],R)$, $x_0(y)\in X$, $0<L_1<1$ be a constant, $a(t,y)$ is continuous function and uniform Hölder continuous in t. Define

$$\mathbf{A}(t)x(t,y)=a(t,y)\frac{\partial^2}{\partial y^2}x(t,y), \tag{1.35}$$

on the domain

$$D(\mathbf{A})=\{x\in X: x,\ \frac{\partial x}{\partial y}\ \text{are absolutely continuous},\ \frac{\partial^2 x}{\partial y^2}\in X,\ x(0)=x(\pi)=0\}.$$

The conditions $(A1)$–$(A3)$ are satisfied and $-\mathbf{A}(t)$ generates a positive evolution system $\mathscr{S}(t,s)$ on X (see [28]). We have $b=t_2=1,\ t_0=s_0=0,\ t_1=\frac{1}{3},\ s_1=\frac{1}{2}$. Put

$$\begin{aligned} x(t)(y)&=x(t,y),\ t\in[0,1], y\in[0,\pi], \\ \mathscr{F}(t,x(t),\mathscr{K}x(t))(y)&=\frac{e^{-t}}{49+e^t}x(t,y)+\int_0^t\frac{1}{50}e^{-s}x(s,y)ds, \\ (\mathscr{K}x(t))(y)&=\int_0^t\frac{1}{50}e^{-s}x(s,y)ds, \\ \gamma_1(t,x(t))(y)&=L_1e^{-(t-\frac{1}{3})}x(t,y), \\ (\mathscr{G}x(t))(y)&=\frac{|x(t,y)|}{7+|x(t,y)|}. \end{aligned} \tag{1.36}$$

The system (1.34) can be transformed into the abstract form (1.23) with nonlocal condition. Now, assume that $x_0(y)\geq 0$ for $y\in[0,\pi]$, and there exists a function $v(t,y)\geq 0$ satisfying

$$v'(t,y)+\mathbf{A}(t)v(t,y)\geq\mathscr{F}(t,v(t,y),\mathscr{K}v(t,y)),\quad t\in\left(0,\frac{1}{3}\right]\cup\left(\frac{1}{2},1\right],\ y\in[0,\pi],$$

$$v(t,0)=v(t,\pi)=0,\quad t\in J,$$

$$v(t, y) \geq L_1 e^{-(t-\frac{1}{3})} v(t, y), \quad y \in (0, \pi),\ t \in \left(\frac{1}{3}, \frac{1}{2}\right],$$
$$v(0, y) \geq \mathscr{G}(v(y)) + x_0(y), \quad y \in [0, \pi].$$

From the above assumptions, we have $\omega_0 = 0$ and $\nu_0 = v(t, y)$ are lower and upper solutions to the system (1.34), which are also lower and upper mild solutions for the problem (1.34). By (1.36), easily we can verify that $(B1)$, $(H2)$ and $(H7)$ hold. Suppose $\{x_n\} \subset [\omega_0(t), \nu_0(t)]$ be a monotone increasing sequence. For $n \leq m$

$$\|\mathscr{F}(t, x_m, \mathscr{K}x_m) - \mathscr{F}(t, x_n, \mathscr{K}x_n)\| \leq \frac{1}{50}(\|x_m - x_n\| + \|\mathscr{K}x_m - \mathscr{K}x_n\|), \text{ hence}$$
$$\mu(\{\mathscr{F}(t, x_n, \mathscr{K}x_n)\}) \leq \frac{1}{50}\Big(\mu(\{x_n\}) + \mu(\mathscr{K}x_n)\Big).$$

Similarly

$$\mu(\{\gamma_1(t, x_n)\}) \leq L_1 \mu(\{x_n\}).$$

Therefore, assumptions $(B2)$, $(H4)$ are satisfied. So, by Theorem 1.5 and Theorem 1.7, we conclude that the minimal and maximal mild solutions for (1.34) exist between the lower solution and upper solutions i.e. in $[0, v]$.

Acknowledgements The first author is supported by Ministry of Human Resource Development, India with grant code "MHR-01-23-200-428" for this work.

References

1. Bainov, D.D., Lakshmikantham, V., Simeonov, P.S.: Theory of Impulsive Differential Equations. Series in Modern Applied Mathematics. World Scientific, Singapore (1989)
2. Banas, J., Goebel, K.: Measure of Noncompactness in Banach Spaces. In: Lecture Notes in Pure and Applied Mathematics, Vol. 60. Marcel Dekker, New York (1980)
3. Benchohra, M., Henderson, J., Ntouyas, S.: Impulsive Differential Equations and Inclusions. Hindawi Publishing Corporation, New York (2006)
4. Byszewski, L.: Theorem about the existence and uniqueness of solutions of a semilinear evolution nonlocal Cauchy problem. J. Math. Anal. Appl. **162**, 494–505 (1991)
5. Chadha, A., Pandey, D.N.: Mild solutions for non-autonomous impulsive semi-linear differential equations with iterated deviating arguments. Electron. J. Differ. Equ. **222**, 1–14 (2015)
6. Chaudhary, R., Pandey, D.N.: Monotone iterative technique for neutral fractional differential equation with infinite delay. Math. Methods Appl. Sci. (2016). Doi: https://doi.org/10.1002/mma.3901
7. Chen, P., Li, Y.: Mixed monotone iterative technique for a class of semilinear impulsive evolution equations in Banach spaces. Nonlinear Anal. **74**, 3578–3588 (2011)
8. Chen, P., Li, Y.: Monotone iterative technique for a class of semilinear evolution equations with nonlocal conditions. Results Math. **63**, 731–744 (2013)
9. Chen, P., Mu, J.: Monotone iterative method for semilinear impulsive evolution equations of mixed type in Banach space. Electron. J. Differ. Equ. **149**, 1–13 (2010)

10. Chen, P., Zhang, X., Li, Y.: Existence of mild solutions to partial differential equations with non-instantaneous impulses. Electron. J. Differ. Equ. **2016**, 1–11 (2016)
11. Deimling, K.: Nonlinear Functional Analysis. Springer, Berlin (1985)
12. Deng, K.: Exponential decay of solutions of semilinear parabolic equations with nonlocal initial conditions. J. Math. Anal. Appl. **179**, 630–637 (1993)
13. Du, S., Lakshmikantam, V.: Monotone iterative technique for differential equation in a Banach space. J. Math. Anal. Appl. **87**, 454–459 (1982)
14. Friedman, A.: Partial Differential Equations. Dover publication, New York (1997)
15. Haloi, R., Pandey, D.N., Bahuguna, D.: Existence, uniqueness and asymptotic stability of solutions to non-autonomous semi-linear differential equations with deviated arguments. Nonlinear Dyn. Syst. Theory. **12**, 179–191 (2012)
16. Heinz, H.P.: On the behaviour of noncompactness with respect to differentiation and integration of vector valued functions. Nonlinear Anal. **7**, 1351–1371 (1983)
17. Hernández, E., O'Regan, D.: On a new class of abstract impulsive differential equations. Proc. Am. Math. Soc. **141**, 1641–1649 (2013)
18. Kamaljeet, Bahuguna, D.: Monotone iterative technique for nonlocal fractional differential equations with finite delay in a Banach space. Electron. J. Qual. Theory Differ. Equ. **9**, 1–16 (2015)
19. Karunanithi, S., Chandrasekaran, S.: Existence result for non-autonomous semilinear integro-differential systems. Int. J. Nonlinear Sci. **13**, 220–227 (2012)
20. Kumar, P., Haloi, R., Bahuguna, D., Pandey, D.N.: Existence of solutions to a new class of abstract non-instantaneous impulsive fractional integro-differential equations. Nonlinear Dyn. Syst. Theory **16**, 73–85 (2016)
21. Kumar, P., Pandey, D.N., Bahuguna, D.: On a new class of abstract impulsive functional differential equations of fractional order. J. Nonlinear Sci. Appl. **7**, 102–114 (2014)
22. Liang, J., Liu, J. H., Xiao, T.J.: Nonlocal impulsive problems for integrodifferential equations. Math. Comput. Model. **49**, 789–804 (2009)
23. Liang, S., Mei, R.: Existence of mild solutions for fractional impulsive neutral evolution equations with nonlocal conditions. Adv. Differ. Equ. (2014). Doi: https://doi.org/10.1186/1687-1847-2014-101
24. Meraj, A., Pandey, D.N.: Monotone iterative technique for non-autonomous semilinear differential equations with nonlocal conditions. Demonstr. Math. **52**, 29–39 (2019)
25. Mu, J.: Monotone iterative technique for fractional evolution equations in Banach spaces. J. Appl. Math. (2011). Doi: https://doi.org/10.1155/2011/767186
26. Mu, J.: Extremal mild solutions for impulsive fractional evolution equations with nonlocal initial conditions. Bound. Value Probl. **71**, 1–12 (2012)
27. Mu, J., Li, Y.: Monotone iterative technique for impulsive fractional evolution equations. J. Inequal. Appl. **125**, 1–12 (2011)
28. Pazy, A.: Semigroup of Linear Operators and Applications to Partial Differential Equations. Springer, New York (1983)
29. Pierri, M., O'Regan, D., Rolnik, V.: Existence of solutions for semi-linear abstract differential equations with not instantaneous impulses. Appl. Math. Comput. **219**, 6743–6749 (2013)
30. Yan, Z.: On solutions of semilinear evolution integro-differential equations with nonlocal conditions. Tamkang J. Math. **40**, 257–269 (2009)
31. Ye, H., Gao, J., Ding, Y.: A generalized Gronwall inequality and its application to a fractional differential equation. J. Math. Anal. Appl. **328**, 1075–1081 (2007)
32. Zhang, X., Li, Y., Chen, P.: Existence of extremal mild solutions for the initial value problem of evolution equations with non-instantaneous impulses. J. Fixed Point Theory Appl. **19**, 3013–3027 (2017)

Chapter 2
An Extrapolated Crank Nicholson VMS-POD Method for Darcy Brinkman Equations

Fatma G. Eroglu and Songul Kaya Merdan

2.1 Introduction

Double diffusive is of great importance in many applications such as oceanography, geology, biology and chemical processes. Although tremendous development of computing power is available, solving Darcy Brinkman equations accurately and efficiently remains a challenge for the computational fluid dynamics community.

The dimensionless form of governing equations of Darcy Brinkman system reads: for the velocity $\mathbf{u} : [0, \tau] \times \Omega \to \mathbb{R}^d$, the pressure $p : [0, \tau] \times \Omega \to \mathbb{R}$, the temperature $T : [0, \tau] \times \Omega \to \mathbb{R}$ and the concentration $C : [0, \tau] \times \Omega \to \mathbb{R}$,

$$
\begin{aligned}
\mathbf{u}_t - 2\nu\nabla\cdot\mathbb{D}\mathbf{u} + (\mathbf{u}\cdot\nabla)\mathbf{u} + Da^{-1}\mathbf{u} + \nabla p &= (\beta_T T + \beta_C C)\mathbf{g} && \text{in } (0,\tau]\times\Omega,\\
\nabla\cdot\mathbf{u} &= 0 && \text{in } (0,\tau]\times\Omega,\\
\mathbf{u} &= \mathbf{0} && \text{in } (0,\tau]\times\partial\Omega,\\
T_t + \mathbf{u}\cdot\nabla T &= \gamma\Delta T && \text{in } (0,\tau]\times\partial\Omega,\\
C_t + \mathbf{u}\cdot\nabla C &= D_c\Delta C && \text{in } (0,\tau]\times\partial\Omega.
\end{aligned}
\tag{2.1}
$$

Here, $\Omega \subset \mathbb{R}^d, d \in \{2, 3\}$ is a confined porous enclosure domain with polygonal boundary $\partial\Omega$. Let Γ_N be a regular open subset of the boundary and $\Gamma_D = \partial\Omega \setminus \Gamma_N$. In addition, in (2.1), the kinematic viscosity is $\nu > 0$, the velocity deformation tensor is $\mathbb{D}\mathbf{u} = (\nabla\mathbf{u} + \nabla\mathbf{u}^T)/2$, the Darcy number is Da, the mass diffusivity is

F. G. Eroglu
Department of Mathematics, Middle East Technical University, Ankara, Turkey

Department of Mathematics, Bartin University, Bartin, Turkey
e-mail: fguler@bartin.edu.tr

S. K. Merdan (✉)
Department of Mathematics, Middle East Technical University, Ankara, Turkey
e-mail: smerdan@metu.edu.tr

© Springer Nature Switzerland AG 2020
J. A. T. Machado et al. (eds.), *Numerical Solutions of Realistic Nonlinear Phenomena*, Nonlinear Systems and Complexity 31, https://doi.org/10.1007/978-3-030-37141-8_2

$D_c > 0$, the thermal diffusivity is $\gamma > 0$, and the gravitational acceleration vector is $\mathbf{g}$, the end time is τ, the thermal and solutal expansion coefficients are β_T and β_C, respectively. The system (2.1) is also equipped with the following initial velocity, temperature, and concentration $\mathbf{u}_0$, C_0, T_0 and suitable boundary conditions.

$$\begin{aligned} \mathbf{u}(0,\mathbf{x}) = \mathbf{u}_0,\ T(0,\mathbf{x}) = T_0,\ C(0,\mathbf{x}) &= C_0 \text{ in } \Omega, \\ T, C = \zeta \qquad\qquad & \text{on } \Gamma_D, \\ \frac{\partial T}{\partial n} = 0, \quad \frac{\partial C}{\partial n} = 0 \qquad & \text{on } \Gamma_N. \end{aligned} \tag{2.2}$$

The dimensionless parameters are given as the Schmidt number Sc, the Prandtl number Pr, the buoyancy ratio N, the thermal and solutal Grashof numbers Gr_T and Gr_C, respectively.

When heat and mass diffuse at various rates, it leads to a complicated fluid motion which is known as double diffusive convection. The detailed derivation of the system (2.1) can be found in [20] and the physical mechanism of double diffusive effects was studied in several works, e.g., [25, 26]. The double diffusive convection was also studied numerically in different flow configurations [3, 5, 10, 14, 15, 19]. The simulation of the system (2.1) by direct numerical simulation (DNS) can be very expensive, and sometimes is not possible due to the wide range of scales. Furthermore, the use of full order methods leads to large algebraic systems and high computational time. These difficulties can be reduced with the emergence of model order reduction method.

The most commonly used reduced order model is the proper orthogonal decomposition (POD). The basic idea of the POD is to use only the most energetic basis functions instead of using billions of basis functions to approximate the solution. This method has been found to be highly efficient for many different types of flow problems. In particular, recent works of [22, 23] with POD have shown that the approach can work well on multiphysics flow problems such as the Boussinesq system for fluids driven by a single potential, and also for magnetohydrodynamics flow [21]. The extension of POD methodology to flows governed by the system (2.1) has been considered in [7, 8].

Despite the widespread use of POD, as mentioned in [11], POD can behave poorly without some stabilization. In particular, the work of [7] reveals that as Rayleigh number increases, convection cells emerge, then the system (2.1) becomes unstable. In this report, we treat the numerical instability of a POD Galerkin method of [7] with the variational multiscale (VMS) method introduced in [13, 16]. Recent works [11, 12] show the efficiency of the VMS-POD in many multiphysics problems such as convection–diffusion–reaction equations and Navier–Stokes equations. For this purpose, we develop the results in [7] by adding a projection-based VMS method to POD method for the velocity, temperature and concentration. The finite element method is considered for space variables and Crank Nicholson time discretization method is considered for time variables. In addition, to obtain a fully linear system at each time level, the nonlinear terms are treated with the extrapolated Crank Nicholson method of Baker's [2].

This work is organized as follows. The weak formulation of Darcy Brinkman system is presented in Sect. 2.2. The basic idea of the POD method and the VMS method is given in Sect. 2.3. The numerical analysis of the VMS-POD method is presented in Sect. 2.4. Numerical experiments are shown in Sect. 2.5 to verify the analytical results and conclusions are given in Sect. 2.6.

2.2 Full Order Model for Darcy Brinkman System

In this study, we consider the standard notations for Sobolev spaces $W^{k,p}(\Omega)$ and Lebesgue spaces $L^p(\Omega)$, $\forall p \in [1, \infty]$, $k \in \mathbb{R}$ c.f. [1]. The norm in $(W^{k,2}(\Omega))^d = (H^k(\Omega))^d$ is denoted by $\|\cdot\|_k$. The $L^p(\Omega)$ norms, for $p \neq 2$ is given by $\|\cdot\|_{L^p}$. If $p = 2$, the $L^2(\Omega)$ space is equipped with the inner product $(\cdot,\cdot)$ and the norm $\|\cdot\|$, respectively. The discrete norms, for $w^n \in H^p(\Omega)$, $n = 0, 1, 2, ..., M$ are denoted by

$$|||w|||_{\infty,p} := \max_{0\le n\le M} \|w^n\|_p \quad |||w|||_{m,p} := \left(\Delta t \sum_{n=0}^{M} \|w^n\|_p^m\right)^{1/m}.$$

The continuous velocity, pressure space and the divergence free spaces are

$$\mathbf{X} := (\mathbf{H}_0^1(\Omega))^d, \quad Q := L_0^2(\Omega), \quad \mathbf{V} := \{\mathbf{v} \in \mathbf{X} : (\nabla\cdot\mathbf{v}, q) = 0,\ \forall q \in Q\},$$

and, the continuous temperature and concentration are

$$W := \{S \in H^1(\Omega) : S = 0 \text{ on } \Gamma_D\}, \quad \Psi := \{\Phi \in H^1(\Omega) : \Phi = 0 \text{ on } \Gamma_D\},$$

The weak formulation of (2.1) reads: Find $\mathbf{u} : (0, \tau] \to \mathbf{X}$, $p : (0, \tau] \to Q$, $T : [0, \tau] \to W$ and $C : [0, \tau] \to \Psi$ satisfying for all $(\mathbf{v}, q, S, \Phi) \in (X, Q, W, \Psi)$.

$$(\mathbf{u}_t, \mathbf{v}) + 2\nu(\mathbb{D}\mathbf{u}, \mathbb{D}\mathbf{v}) + b_1(\mathbf{u}, \mathbf{u}, \mathbf{v}) + (Da^{-1}\mathbf{u}, \mathbf{v}) - (p, \nabla\cdot\mathbf{v}) = \beta_T(\mathbf{g}T, \mathbf{v}) + \beta_C(\mathbf{g}C, \mathbf{v}), \tag{2.3}$$

$$(T_t, S) + b_2(\mathbf{u}, T, S) + \gamma(\nabla T, \nabla S) = 0, \tag{2.4}$$

$$(C_t, \Phi) + b_3(\mathbf{u}, C, \Phi) + D_c(\nabla C, \nabla\Phi) = 0, \tag{2.5}$$

where $b_i(w_1, w_2, w_3) = \frac{1}{2}(((w_1\cdot\nabla)w_2, w_3) - ((w_1\cdot\nabla)w_3, w_2))$, $\forall i = 1, 2, 3$ defines the skew-symmetric forms of the convective terms for each variables.

The following bounds are used in the error analysis.

Lemma 2.1 *The trilinear skew-symmetric forms satisfy the following bounds*

$$b_i(u, v, v) = 0,$$
$$b_i(u, v, w) \leq K\sqrt{\|u\|\|\nabla u\|}\|\nabla v\|\|\nabla w\|,$$
$$b_i(u, v, w) \leq K\|\nabla u\|\|\nabla v\|\|\nabla w\|,$$

for generic constant $K = K(\Omega)$.

Proof See, e.g., [9, 17] for a proof.

Lemma 2.2 *Let* $w(t, \mathbf{x})$ *be a function and* $t^{n/2} = \frac{t^{n+1}+t^n}{2}$. *Then, for all* $w, w_t, w_{tt}, w_{ttt} \in C^0(0, \tau, L^2(\Omega))$ *and for all* $t^* \in (t_0, \tau)$, *the following inequalities hold:*

$$\left\|\frac{w(t^{n+1}) + w(t^n)}{\Delta t}\right\| \leq K\|w_t(t^*)\|, \tag{2.6}$$

$$\left\|\frac{w(t^{n+1}) + w(t^n)}{2} - w(t^{n/2})\right\| \leq K\Delta t^2\|w_{tt}(t^*)\|, \tag{2.7}$$

$$\left\|\frac{3w(t^n)}{2} - \frac{w(t^{n-1})}{2} - w(t^{n/2})\right\| \leq K\Delta t^2\|w_{tt}(t^*)\|, \tag{2.8}$$

$$\left\|\frac{w(t^{n+1}) + w(t^n)}{\Delta t} - w_t(t^{n/2})\right\| \leq K\Delta t^2\|w_{ttt}(t^*)\|. \tag{2.9}$$

Here, $w(x, t^n)$ *is denoted by* $w(t^n)$.

Proof This can be proved by using Taylor series expansion of $w(t, \mathbf{x})$.

Let τ^h be a triangulation of Ω and $\mathbf{X}_h \subset \mathbf{X}$, $Q_h \subset Q$, $W_h \subset W$ and $\Psi_h \subset \Psi$ be conforming finite element spaces. It is assumed that the pair $(\mathbf{X}_h, Q_h)$ provides the discrete inf-sup condition, see [9]. For simplicity, it is also be assumed that the finite element spaces $\mathbf{X}_h$, W_h, Ψ_h are composed of piecewise polynomials of degree at most m and Q_h is composed of piecewise polynomials of degree at most $m - 1$. In addition, we assume that the spaces satisfy standard interpolation estimates. We define the discretely divergence free space $\mathbf{V}_h$ for $(\mathbf{X}_h, Q_h)$:

$$\mathbf{V}_h = \{\mathbf{v}_h \in \mathbf{X}_h : (\nabla \cdot \mathbf{v}_h, q_h) = 0, \forall q_h \in Q_h\}. \tag{2.10}$$

The inf-sup condition implies that the space $\mathbf{V}_h$ is a closed subspace of $\mathbf{X}_h$ and the formulation above involving $\mathbf{X}_h$ and Q_h is equivalent to the following $\mathbf{V}_h$ formulation: Hence, the variational formulation of (2.3) reads as: Find $\mathbf{u}_h : [0, \tau] \to \mathbf{V}_h$, $T_h : [0, \tau] \to W_h$, $C_h : [0, \tau] \to \Psi_h$ satisfying

$$\begin{aligned} (\mathbf{u}_{h,t}, \mathbf{v}_h) + 2\nu(\mathbb{D}\mathbf{u}_h, \mathbb{D}\mathbf{v}_h) + b_1(\mathbf{u}_h, \mathbf{u}_h, \mathbf{v}_h) + (Da^{-1}\mathbf{u}_h, \mathbf{v}_h) &= \beta_T(\mathbf{g}T_h, \mathbf{v}_h) \\ &\quad + \beta_C(\mathbf{g}C_h, \mathbf{v}_h), \\ (T_{h,t}, S_h) + b_2(\mathbf{u}_h, T_h, S_h) + \gamma(\nabla T_h, \nabla S_h) &= 0, \\ (C_{h,t}, \Phi_h) + b_3(\mathbf{u}_h, C_h, \Phi_h) + D_c(\nabla C_h, \nabla \Phi_h) &= 0, \end{aligned} \tag{2.11}$$

for all $(\mathbf{v}_h, S_h, \Phi_h) \in (\mathbf{V}_h, W_h, \psi_h)$.

2.3 Reduced Order Modelling with POD

We consider the snapshots $\{\mathbf{u}(\cdot, t_i)\}_{i=1}^{M_1}$, $\{T(\cdot, t_i)\}_{i=1}^{M_2}$, $\{C(\cdot, t_i)\}_{i=1}^{M_3}$ at different M_1, M_2, M_3 instances—for the velocity, temperature and concentration, respectively. These snapshots come from DNS obtained by finite element spatial discretization. The main purpose of the POD is to find low dimensional bases for velocity, temperature, concentration by solving the minimization problems of

$$\underset{\boldsymbol{\psi}_1, \boldsymbol{\psi}_2, \ldots, \boldsymbol{\psi}_{r_1}}{\arg\min} \frac{1}{M_1} \sum_{k=1}^{M_1} \left\| \mathbf{u}(\cdot, t_k) - \sum_{i=1}^{r_1} (\mathbf{u}(\cdot, t_k), \boldsymbol{\psi}_i(\cdot)) \boldsymbol{\psi}_i(\cdot) \right\|^2, \quad \text{subject to } (\boldsymbol{\psi}_i, \boldsymbol{\psi}_j) = \delta_{ij}, \tag{2.12}$$

$$\underset{\phi_1, \phi_2, \ldots, \phi_{r_2}}{\arg\min} \frac{1}{M_2} \sum_{k=1}^{M_2} \left\| T(\cdot, t_k) - \sum_{i=1}^{r_2} (T(\cdot, t_k), \phi_i(\cdot)) \phi_i(\cdot) \right\|^2, \quad \text{subject to } (\boldsymbol{\phi}_i, \boldsymbol{\phi}_j) = \delta_{ij}, \tag{2.13}$$

$$\underset{\eta_1, \eta_2, \ldots, \eta_{r_1}}{\arg\min} \frac{1}{M_3} \sum_{k=1}^{M_3} \left\| C(\cdot, t_k) - \sum_{i=1}^{r_3} (C(\cdot, t_k), \eta_i(\cdot)) \eta_i(\cdot) \right\|^2, \quad \text{subject to } (\eta_i, \eta_j) = \delta_{ij}, \tag{2.14}$$

for all $1 \leq i, j \leq r_s$ such that $r_s << N_s$, $\forall s = 1, 2, 3$. The discretization of the problem (2.12)–(2.14) leads to large eigenvalue problems. The method of [24] allows that eigenvalue problems are transformed into much smaller and traceable problems. When the eigenvalue problem is solved, the POD basis functions are calculated as

$$\boldsymbol{\psi}_k(\cdot) = \frac{1}{\sqrt{\lambda_k}} \sum_{i=1}^{M_1} (\mathbf{v}_k)_i \mathbf{u}(\cdot, t_i), \quad 1 \leq k \leq r_1, \tag{2.15}$$

$$\phi_k(\cdot) = \frac{1}{\sqrt{\mu_k}} \sum_{i=1}^{M_2} (S_k)_i T(\cdot, t_i), \quad 1 \le k \le r_2, \tag{2.16}$$

$$\eta_k(\cdot) = \frac{1}{\sqrt{\gamma_k}} \sum_{i=1}^{M_3} (\Phi_k)_i C(\cdot, t_i), \quad 1 \le k \le r_3, \tag{2.17}$$

where λ_k, μ_k, ξ_k denote the eigenvalues of the snapshots correlation matrices and $(\mathbf{v}_k)_i$, $(S_k)_i$ and $(\Phi_k)_i$ denote the ith components of the eigenvectors $\mathbf{v}_k$, S_k, Φ_k. Since all eigenvalues are sorted in descending order, the basis functions $\{\boldsymbol{\psi}_k\}_{k=1}^{r_1}$, $\{\phi_k\}_{k=1}^{r_2}$ and $\{\eta_k\}_{k=1}^{r_3}$ correspond to the first r_1, r_2 and r_3 largest eigenvalues, respectively. Let $\mathbf{X}_r$, W_r and Ψ_r be the POD spaces spanned by POD basis functions:

$$\mathbf{X}_r = span\{\boldsymbol{\psi}_k\}_{k=1}^{r_1}, \quad W_r = span\{\phi_k\}_{k=1}^{r_2}, \quad \Psi_r = span\{\eta_k\}_{k=1}^{r_3}.$$

Then the POD solutions of (2.11) are constructed by writing

$$\mathbf{u}_r(t, \mathbf{x}) = \sum_{k=1}^{r_1} a_k(t)\boldsymbol{\psi}_k(\mathbf{x}), \; T_r(t, \mathbf{x}) = \sum_{k=1}^{r_2} b_k(t)\phi_k(\mathbf{x}), \; C_r(t, \mathbf{x}) = \sum_{k=1}^{r_3} c_k(t)\eta_k(\mathbf{x}),$$

where $\{a_k(t)\}_{k=1}^{r_1}$, $\{b_k(t)\}_{k=1}^{r_2}$, $\{c_k(t)\}_{k=1}^{r_3}$ sought time varying coefficients representing the POD Galerkin trajectories. The L^2 projection operators, which we will use in error analysis, are given as

$$P_{u,r} : L^2 \to \mathbf{X}_r, \quad P_{T,r} : L^2 \to W_r, \quad P_{C,r} : L^2 \to \Psi_r,$$

which are defined by

$$(\mathbf{u} - P_{u,r}\mathbf{u}, \mathbf{v}_r) = 0, \quad (T - P_{T,r}T, S_r) = 0, \quad (C - P_{C,r}C, \zeta_r) = 0, \tag{2.18}$$

for all $(\mathbf{v}_r, S_r, \zeta_r) \in (\mathbf{X}_r, W_r, \Psi_r)$. We now state the L^2 projection (2.18) error estimations in general. For a detailed derivation of them, the reader is referred to [6, 12].

Lemma 2.3 *Let $w \in L^\infty(0, k; H^{m+1}(\Omega))$ be fulfilled and its L^2 projection $P_{w,r} : L^2 \to X_r^w$, where X_r^w is the POD space. Let $\| \cdot \|_2$ be matrix 2-norm. For any $w^n \in H_0^1(\Omega)$ the following inequalities are provided:*

$$\frac{1}{M} \sum_{n=0}^{M} \|w^n - P_{w,r}w^n\|^2 \le K\Big(h^{2m+2} + \sum_{i=r_w+1}^{d} \lambda_w\Big), \tag{2.19}$$

$$\frac{1}{M} \sum_{n=0}^{M} \|\nabla(w^n - P_{w,r}w^n)\|^2 \le K\Big(h^{2m} + \|S_{w,r}\|_2 h^{2m+2} + \varepsilon_w^2\Big), \tag{2.20}$$

where $(S_{w,r})_{i,j} = \int_{\Omega} \nabla \boldsymbol{\psi}_j^w \cdot \nabla \boldsymbol{\psi}_i^w$ *is the POD stiffness matrices,* $\varepsilon_w = \sqrt{\sum_{i=r_w+1}^{d} \|\boldsymbol{\psi}_i^w\|_1^2 \lambda_i^w}$ *is the POD contribution,* $\{\boldsymbol{\psi}_i^w\}_{i=1}^{r_w}$ *are the POD basis functions and* $\{\lambda_i^w\}_{i=1}^{r_w}$ *are the corresponding eigenvalues.*

We now assume that the similar estimations in Lemma 2.3 are held also for the single term, which is logical and valid.

Assumption 1 *We assume that the following estimations are held:*

$$\|w^n - \tilde{w}^n\|^2 \leq K\Big(h^{2m+2} + \sum_{i=r_w+1}^{d} \lambda_w\Big), \tag{2.21}$$

$$\|\nabla(w^n - \tilde{w}^n)\|^2 \leq K\Big(h^{2m} + \|S_{w,r}\|_2 h^{2m+2} + \varepsilon_w^2\Big). \tag{2.22}$$

The POD formulation of the Darcy Brinkman double diffusion system for selected $\mathbf{u}_0 \in (L^2(\Omega))^d$, $T_0, C_0 \in L^2(\Omega)$, $\mathbf{g} \in L^\infty(0,k; L^p(\Omega))$: Find $\mathbf{u}_r : [0,\tau] \to \mathbf{X}_r$, $T_r : [0,\tau] \to W_r$, $C_r : [0,\tau] \to \Psi_r$ satisfying

$$\begin{aligned}
(\mathbf{u}_{r,t}, \mathbf{v}_r) + 2\nu(\mathbb{D}\mathbf{u}_r, \mathbb{D}\mathbf{v}_r) + b_1(\mathbf{u}_r, \mathbf{u}_r, \mathbf{v}_r) + (Da^{-1}\mathbf{u}_r, \mathbf{v}_r) &= \beta_T(\mathbf{g}T_r, \mathbf{v}_r) \\
&\quad + \beta_C(\mathbf{g}C_r, \mathbf{v}_r), \\
(T_{r,t}, S_r) + b_2(\mathbf{u}_r, T_r, S_r) + \gamma(\nabla T_r, \nabla S_r) &= 0, \\
(C_{r,t}, \Phi_r) + b_3(\mathbf{u}_r, C_r, \Phi_r) + D_c(\nabla C_r, \nabla \Phi_r) &= 0,
\end{aligned} \tag{2.23}$$

for every $(\mathbf{v}_r, S_r, \Phi_r), \in (\mathbf{X}_r, W_r, \Psi_r)$. Although POD is a widely used successful reduced order model, it is not effective enough in case of high Reynolds number [6, 11, 12]. In such a case, we use the VMS method to eliminate the oscillation and stabilize the convective terms. In other words, we add artificial diffusions to the smaller R_1, R_2, R_3 velocity, temperature, concentration modes affecting only small scales. Thus, the following spaces are required for the construction of the VMS method. For $R_1 < r_1$, $R_2 < r_2$ and $R_3 < r_3$

$$\begin{aligned}
\mathbf{X}_R &= span\{\boldsymbol{\psi}_k\}_{k=1}^{R_1}, \\
W_R &= span\{\boldsymbol{\phi}_k\}_{k=1}^{R_2}, \\
\Psi_R &= span\{\eta_k\}_{k=1}^{R_3}.
\end{aligned}$$

and

$$\begin{aligned}
\mathbf{L}_{R,\mathbf{u}} &= \nabla \mathbf{X}_R := \{\nabla \boldsymbol{\psi}_k\}_{k=1}^{R_1}, \\
\mathbf{L}_{R,T} &= \nabla W_R := \{\nabla \boldsymbol{\phi}_k\}_{k=1}^{R_2}, \\
\mathbf{L}_{R,C} &= \nabla \Psi_R := \{\nabla \eta_k\}_{k=1}^{R_3},
\end{aligned}$$

where R_1, R_2 and R_3 are the VMS modes numbers. According to VMS framework, the following relation holds:

$$\mathbf{X}_R \subset \mathbf{X}_r \subset \mathbf{V}_h \subset \mathbf{X}.$$

We note that similar relations are also satisfied for the temperature and concentration spaces. The L^2 projection operators $P_{u,R} : L^2 \to \mathbf{L}_{R,\mathbf{u}}$, $P_{T,R} : L^2 \to \mathbf{L}_{R,T}$, $P_{C,R} : L^2 \to \mathbf{L}_{R,C}$ are defined by

$$\begin{aligned}
(\mathbf{u} - P_{u,R}\mathbf{u}, \mathbf{v}_R) &= 0,\\
(T - P_{T,R}T, S_R) &= 0,\\
(C - P_{C,R}C, \zeta_R) &= 0,
\end{aligned} \tag{2.24}$$

for all $(\mathbf{v}_R, S_R, \zeta_R) \in (\mathbf{L}_{R,\mathbf{u}}, \mathbf{L}_{R,T}, \mathbf{L}_{R,C})$. Thus, the VMS-POD solution of (2.1) based on the Crank Nicholson time discretization becomes: Find $\mathbf{u}_r : [0, \tau] \to \mathbf{X}_r$, $\quad T_r : [0, \tau] \to W_r$, $\quad C_r : [0, \tau] \to \Psi_r$

$$\begin{aligned}
&\left(\frac{\mathbf{u}_r^{n+1} - \mathbf{u}_r^n}{\Delta t}, \mathbf{v}_r\right) + 2\nu(\mathbb{D}\mathbf{u}_r^{n/2}, \mathbb{D}\mathbf{v}_r) + b_1(\mathscr{X}(\mathbf{u}_r^n), \mathbf{u}_r^{n/2}, \mathbf{v}_r)\\
&\quad+\alpha_1\Big((I - P_{\mathbf{u},R})\mathbb{D}\mathbf{u}_r^{n/2}, (I - P_{\mathbf{u},R})\mathbb{D}\mathbf{v}_r\Big) + Da^{-1}(\mathbf{u}_r^{n/2}, \mathbf{v}_r)\\
&= \beta_T(\mathbf{g}T_r^{n/2}, \mathbf{v}_r) + \beta_C(\mathbf{g}C_r^{n/2}, \mathbf{v}_r),
\end{aligned} \tag{2.25}$$

$$\begin{aligned}
&\left(\frac{T_r^{n+1} - T_r^n}{\Delta t}, S_r\right) + \gamma(\nabla T_r^{n/2}, \nabla S_r) + b_2(\mathscr{X}(\mathbf{u}_r^n), T_r^{n/2}, S_r)\\
&\quad+\alpha_2\Big((I - P_{T,R})\nabla T_r^{n/2}, (I - P_{T,R})\nabla S_r\Big) = 0,
\end{aligned} \tag{2.26}$$

$$\begin{aligned}
&\left(\frac{C_r^{n+1} - C_r^n}{\Delta t}, \varsigma_r\right) + D_c(\nabla C_r^{n/2}, \nabla \varsigma_r) + b_3(\mathscr{X}(\mathbf{u}_r^n), C_r^{n/2}, \varsigma_r)\\
&\quad+\alpha_3\Big((I - P_{C,R})\nabla C_r^{n/2}, (I - P_{C,R})\nabla \varsigma_r\Big) = 0,
\end{aligned} \tag{2.27}$$

for all $(\mathbf{v}_r, S_r, \Phi_r), \in (\mathbf{X}_r, W_r, \Psi_r)$ where $P_{\mathbf{u},R}$, $P_{T,R}$ and $P_{C,R}$ are the L^2 projections into $(\mathbf{X}_R, W_R, \Psi_R)$ and

$$\mathbf{u}_r^{n/2} = \frac{\mathbf{u}_r^{n+1} + \mathbf{u}_r^n}{2},\ T_r^{n/2} = \frac{T_r^{n+1} + T_r^n}{2},$$

$$C_r^{n/2} = \frac{C_r^{n+1} + C_r^n}{2},\ \mathscr{X}(\mathbf{u}_r^n) = \frac{3}{2}\mathbf{u}_r^n - \frac{1}{2}\mathbf{u}_r^{n-1}.$$

Note that, since linear extrapolations are utilized for each fluid variables in (2.25)–(2.27), the solution of the system needs one linear system per time.

2.4 Error Estimates

In this section, we perform the numerical analysis for the solutions of (2.25)–(2.27). We first perform the stability analysis.

Lemma 2.4 *The VMS-POD approximation (2.25)–(2.27) is unconditionally stable in the following sense: for any $\Delta t > 0$, one has*

$$\|\mathbf{u}_r^M\| + 4\nu\Delta t \sum_{n=0}^{M-1} \|\mathbb{D}\mathbf{u}_r^{n/2}\|^2 + 2Da^{-1}\Delta t \sum_{n=0}^{M-1} \|\mathbf{u}_r^{n/2}\|^2$$
$$+2\alpha_1\Delta t \sum_{n=0}^{M-1} \|(I - P_{\mathbf{u},R})\mathbb{D}\mathbf{u}_r^{n/2}\|^2$$
$$\leq \|\mathbf{u}_r^0\|^2 + K\|\mathbf{g}\|_\infty^2(\beta_T^2\nu^{-1}\gamma^{-1}\|T_r^0\|^2 + \beta_C^2 Da D_c^{-1}\|C_r^0\|^2), \tag{2.28}$$

$$\|T_r^M\|^2 + 2\gamma\Delta t \sum_{n=0}^{M-1} \|\nabla T_r^{n/2}\| + 2\alpha_2\Delta t \sum_{n=0}^{M-1} \|(I - P_{T,R})\nabla T_r^{n/2}\|^2$$
$$\leq \|T_r^0\|^2, \tag{2.29}$$

$$\|C_r^M\|^2 + 2D_c\Delta t \sum_{n=0}^{M-1} \|\nabla C_r^{n/2}\| + 2\alpha_3\Delta t \sum_{n=0}^{M-1} \|(I - P_{C,R})\nabla C_r^{n/2}\|^2$$
$$\leq \|C_r^0\|^2. \tag{2.30}$$

Proof Setting $S_r = T_r^{n/2} = \dfrac{T_r^{n+1} + T_r^n}{2}$ in (2.26), and using the skew symmetry, and summing from $n = 0$ to $M - 1$ gives

$$\|T_r^M\|^2 + 2\gamma\Delta t \textstyle\sum_{n=0}^{M-1} \|\nabla T_r^{n/2}\|$$
$$+2\alpha_2\Delta t \textstyle\sum_{n=0}^{M-1} \|(I - P_{T,R})\nabla T_r^{n/2}\|^2 \leq \|T_r^0\|^2. \tag{2.31}$$

In a similar manner, choosing $\varsigma_r = C_r^{n/2}$ in (2.27) yields

$$\|C_r^M\|^2 + 2D_c\Delta t \textstyle\sum_{n=0}^{M-1} \|\nabla C_r^{n/2}\|$$
$$+2\alpha_3\Delta t \textstyle\sum_{n=0}^{M-1} \|(I - P_{C,R})\nabla C_r^{n/2}\|^2 \leq \|C_r^0\|^2. \tag{2.32}$$

Letting $\mathbf{v}_r = \mathbf{u}_r^{n/2}$ and utilizing Lemma 2.1 in (2.25), we get

$$\begin{aligned}\|\mathbf{u}_r^{n+1}\| + 4\nu\Delta t\|\mathbb{D}\mathbf{u}_r^{n/2}\|^2 + 2\alpha_1\Delta t\|(I - P_{u,R})\mathbb{D}\mathbf{u}_r^{n/2}\|^2 \\ +2Da^{-1}\Delta t\|\mathbf{u}_r^{n/2}\|^2 = \|\mathbf{u}_r^n\|^2 + 2\beta_T\Delta t(\mathbf{g}T_r^{n/2}, \mathbf{u}_r^{n/2}) \\ +2\beta_C\Delta t(\mathbf{g}C_r^{n/2}, \mathbf{u}_r^{n/2}). \qquad (2.33)\end{aligned}$$

Performing Cauchy–Schwarz and Young's inequalities, and summing from $n = 0$ to $M - 1$, we have

$$\begin{aligned}&\|\mathbf{u}_r^M\| + 4\nu\Delta t\sum_{n=0}^{M-1}\|\mathbb{D}\mathbf{u}_r^{n/2}\|^2 + 2\alpha_1\Delta t\sum_{n=0}^{M-1}\|(I - P_{u,R})\mathbb{D}\mathbf{u}_r^{n/2}\|^2 \\ &+2Da^{-1}\Delta t\sum_{n=0}^{M-1}\|\mathbf{u}_r^{n/2}\|^2 \leq \|\mathbf{u}_r^0\|^2 \\ &+K\|\mathbf{g}\|_\infty^2\left(\beta_T^2\nu^{-1}\Delta t\sum_{n=0}^{M-1}\|\nabla T_r^{n/2}\| + \beta_C^2 Da\Delta t\sum_{n=0}^{M-1}\|\nabla C_r^{n/2}\|^2\right). \qquad (2.34)\end{aligned}$$

Substituting (2.31) and (2.32) in (2.34) produces the stated result (2.28).

Now, we consider the error analysis of VMS-POD.

Theorem 2.1 (Error Estimation) *Let regularity assumptions* $\mathbf{u}, T, C \in L^\infty(0, \tau; H^{m+1})$*,* $p \in L^\infty(0, \tau; H^m)$ *hold. Then, for a sufficiently small* Δt*, the error satisfies*

$$\begin{aligned}&\|\mathbf{u}^M - \mathbf{u}_r^M\|^2 + \|T^M - T_r^M\|^2 + \|C^M - C_r^M\|^2 \\ &\leq K\Bigg(1 + h^{2m} + (\Delta t)^4 + (1 + \|S_{u,r}\|_2 + \|S_{u,R}\|_2 \\ &+\|S_{T,r}\|_2 + \|S_{T,R}\|_2 + \|S_{C,r}\|_2 + \|S_{C,R}\|_2)h^{2m+2} \\ &+ \sum_{i=r_1+1}^{d}(\|\boldsymbol{\psi}_i\|_1^2 + 1)\lambda_i + \sum_{i=r_2+1}^{d}(\|\phi_i\|_1^2 + 1)\mu_i + \sum_{i=r_3+1}^{d}(\|\eta_i\|_1^2 + 1)\xi_i \\ &+ \sum_{i=R_1+1}^{d}\|\boldsymbol{\psi}_i\|_1^2\lambda_i + \sum_{i=R_2+1}^{d}\|\phi_i\|_1^2\mu_i + \sum_{i=R_3+1}^{d}\|\eta_i\|_1^2\xi_i\Bigg). \qquad (2.35)\end{aligned}$$

Proof At time $t^{n/2}$, subtracting from (2.3), (2.4), (2.5) to (2.25), (2.26), (2.27) at time $t^{n/2}$, respectively, we get

$$\begin{aligned}
&\left(\mathbf{u}_t^{n/2} - \frac{\mathbf{u}_r^{n+1} - \mathbf{u}_r^n}{\Delta t}, \mathbf{v}_r\right) + 2\nu(\mathbb{D}(\mathbf{u}^{n/2} - \mathbf{u}_r^{n/2}), \mathbb{D}\mathbf{v}_r) + b_1(\mathbf{u}^{n/2}, \mathbf{u}^{n/2}, \mathbf{v}_r) \\
&\quad - b_1(\mathscr{X}(\mathbf{u}_r^n), \mathbf{u}_r^{n/2}, \mathbf{v}_r) + (Da^{-1}(\mathbf{u}^{n/2} - \mathbf{u}_r^{n/2}), \mathbf{v}_r) - (p^{n+1}, \nabla \cdot \mathbf{v}_r) \\
&\quad + \alpha_1\Big((I - P_{\mathbf{u},R})\mathbb{D}(\mathbf{u}^{n/2} - \mathbf{u}_r^{n/2}), (I - P_{\mathbf{u},R})\mathbb{D}\mathbf{v}_r\Big) \\
&= \beta_T(\mathbf{g}(T^{n/2} - T_r^{n/2}), \mathbf{v}_r) + \beta_C(\mathbf{g}(C^{n/2} - C_r^{n/2}), \mathbf{v}_r) \\
&\quad + \alpha_1\Big((I - P_{\mathbf{u},R})\mathbb{D}\mathbf{u}^{n/2}, (I - P_{\mathbf{u},R})\mathbb{D}\mathbf{v}_r\Big),
\end{aligned} \tag{2.36}$$

$$\begin{aligned}
&\left(T_t^{n/2} - \frac{T_r^{n+1} - T_r^n}{\Delta t}, S_r\right) + \gamma(\nabla(T^{n/2} - T_r^{n/2}), \nabla S_r) + b_2(\mathbf{u}^{n/2}, T^{n/2}, S_r) \\
&\quad - b_2(\mathscr{X}(\mathbf{u}_r^n), T_r^{n/2}, S_r) + \alpha_2\Big((I - P_{T,R})\nabla(T^{n/2} - T_r^{n/2}), (I - P_{T,R})\nabla S_r\Big) \\
&= \alpha_2\Big((I - P_{T,R})\nabla T^{n/2}, (I - P_{T,R})\nabla S_r\Big),
\end{aligned} \tag{2.37}$$

$$\begin{aligned}
&\left(C_t^{n/2} - \frac{C_r^{n+1} - C_r^n}{\Delta t}, \Phi_r\right) + D_c(\nabla(C^{n/2} - C_r^{n/2}), \nabla \Phi_r) + b_3(\mathbf{u}^{n/2}, C^{n/2}, \Phi_r) \\
&\quad - b_3(\mathscr{X}(\mathbf{u}_r^n), C_r^{n/2}, \Phi_r) + \alpha_3\Big((I - P_{C,R})\nabla(C^{n/2} - C_r^{n/2}), (I - P_{C,R})\nabla \varsigma_r\Big) \\
&= \alpha_3\Big((I - P_{C,R})\nabla C^{n/2}, (I - P_{C,R})\nabla \varsigma_r\Big).
\end{aligned} \tag{2.38}$$

for all $(\mathbf{v}_r, S_r, \Phi_r) \in (\mathbf{X}_r, W_r, \Psi_r)$. We use the following notations for the decomposition of the errors.

$$\begin{aligned}
\boldsymbol{\eta}_{\mathbf{u}}^n &:= \mathbf{u}^n - \tilde{\mathbf{u}}^n, & \boldsymbol{\phi}_{\mathbf{u},r}^n &:= \mathbf{u}_r^n - \tilde{\mathbf{u}}^n, \\
\eta_T^n &= T^n - \tilde{T}^n, & \phi_{T,r}^n &:= T_r^n - \tilde{T}^n, \\
\eta_C^n &= C^n - \tilde{C}^n, & \phi_{C,r}^n &:= C_r^n - \tilde{C}^n,
\end{aligned} \tag{2.39}$$

where $\tilde{\mathbf{u}}^n$, $\tilde{T}^n$, $\tilde{C}^n$ are L^2 projections of $\mathbf{u}^n$, T^n, C^n in $\mathbf{X}_r$, W_r, Ψ_r at time t^n, respectively. Hence the errors can be denoted by

$$\mathbf{e}_{\mathbf{u},r}^n = \boldsymbol{\eta}_{\mathbf{u}}^n - \boldsymbol{\phi}_{\mathbf{u},r}^n, \quad e_{T,r}^n := \eta_T^n - \phi_{T,r}^n, \quad e_{C,r}^n := \eta_C^n - \phi_{C,r}^n. \tag{2.40}$$

We first derive the error estimation for the temperature. To do that, the error equation for the temperature is rewritten as

$$\begin{aligned}
&\left(\frac{T(t^{n+1})-T(t^n)}{\Delta t}-\frac{T_r^{n+1}-T_r^n}{\Delta t}, S_r\right)+\gamma(\nabla(T(t^{n/2})-T_r^{n/2}), \nabla S_r)\\
&\quad+b_2(\mathbf{u}(t^{n/2}), T(t^{n/2}), S_r)-b_2(\mathscr{X}(\mathbf{u}_r^n), T_r^{n/2}, S_r)\\
&\quad+\alpha_2\Big((I-P_{T,R})\nabla(T(t^{n/2})-T_r^{n/2}), (I-P_{T,R})\nabla S_r\Big)\\
&\quad+\left(T_t(t^{n/2})-\frac{T(t^{n+1})-T(t^n)}{\Delta t}, S_r\right)\\
&=\alpha_2\Big((I-P_{T,R})\nabla T(t^{n/2}), (I-P_{T,R})\nabla S_r\Big). \qquad (2.41)
\end{aligned}$$

Adding and subtracting terms

$$\gamma\left(\nabla\left(\frac{T(t^{n+1})+T(t^n)}{\Delta t}\right), \nabla S_r\right)+\alpha_2\Big((I-P_{T,R})\nabla\left(\frac{T(t^{n+1})+T(t^n)}{\Delta t}\right),\\ (I-P_{T,R})\nabla S_r\Big)$$

to (2.41) and utilizing (2.39) and setting $S_r=\phi_{T,r}^{n/2}$ in (2.41) gives

$$\begin{aligned}
&\left(\frac{\phi_{T,r}^{n+1}-\phi_{T,r}^n}{\Delta t}, \phi_{T,r}^{n+1}\right)+\gamma\|\nabla\phi_{T,r}^{n/2}\|^2+\alpha_2\|(I-P_{T,R})\nabla\phi_{T,r}^{n/2}\|^2\\
&\leq\left|\left(\frac{\eta_T^{n+1}-\eta_T^n}{\Delta t}, \phi_{T,r}^{n+1}\right)\right|+\gamma|(\nabla\eta_T^{n/2}, \nabla\phi_{T,r}^{n/2})|\\
&\quad+\gamma\left|\left(\nabla\left(\frac{T(t^{n+1})+T(t^n)}{\Delta t}-T(t^{n/2})\right), \nabla\phi_{T,r}^{n/2}\right)\right|^2\\
&\quad+|b_2(\mathbf{u}(t^{n/2}), T(t^{n/2}), \phi_{T,r}^{n/2})-b_2(\mathscr{X}(\mathbf{u}_r^n), T_r^{n/2}, \phi_{T,r}^{n/2})|\\
&\quad+\alpha_2\left|\Big((I-P_{T,R})\nabla\eta_T^{n/2}, (I-P_{T,R})\nabla\phi_{T,r}^{n/2}\Big)\right|\\
&\quad+\left|\left(\frac{T(t^{n+1})-T(t^n)}{\Delta t}-T_t(t^{n/2})\right), \phi_{T,r}^{n/2})\right|\\
&\quad+\alpha_2\left|\Big((I-P_{T,R})\nabla T(t^{n/2}), (I-P_{T,R})\nabla S_r\Big)\right|\\
&\quad+\alpha_2\left|\Big((I-P_{T,R})\nabla\left(\frac{T(t^{n+1})+T(t^n)}{\Delta t}-T(t^{n/2})\right), (I-P_{T,R})\nabla S_r\Big)\right|. \qquad (2.42)
\end{aligned}$$

Using the fact that $(\eta_T^{n+1}, \phi_{T,r}^{n+1}) = 0$, and $(\eta_T^n, \phi_{T,r}^{n+1}) = 0$ from the definition of L^2 projection in (2.42), we get

$$
\begin{aligned}
&\frac{1}{2\Delta t}\|\phi_{T,r}^{n+1}\|^2 + \gamma\|\nabla\phi_{T,r}^{n/2}\|^2 + \alpha_2\|(I - P_{T,R})\nabla\phi_{T,r}^{n/2}\|^2 \\
&\le \frac{1}{2\Delta t}\|\phi_{T,r}^n\| + \gamma|(\nabla\eta_T^{n/2}, \nabla\phi_{T,r}^{n/2})| + \gamma\left|\left(\nabla\left(\frac{T(t^{n+1}) + T(t^n)}{\Delta t} - T(t^{n/2})\right), \nabla\phi_{T,r}^{n/2}\right)\right|^2 \\
&+ \left|b_2(\mathbf{u}(t^{n/2}), T(t^{n/2}), \phi_{T,r}^{n/2}) - b_2(\mathscr{X}(\mathbf{u}_r^n), T_r^{n/2}, \phi_{T,r}^{n/2})\right| \\
&+\alpha_2\left|\left((I - P_{T,R})\nabla\eta_T^{n/2}, (I - P_{T,R})\nabla\phi_{T,r}^{n/2}\right)\right| \\
&+\alpha_2\left|\left((I - P_{T,R})\nabla T(t^{n/2}), (I - P_{T,R})\nabla\phi_{T,r}^{n/2}\right)\right| \\
&+\alpha_2\left|\left((I - P_{T,R})\nabla(\frac{T(t^{n+1}) + T(t^n)}{\Delta t} - T(t^{n/2})), (I - P_{T,R})\nabla\phi_{T,r}^{n/2}\right)\right| \\
&+\left|\left(\frac{T(t^{n+1}) - T(t^n)}{\Delta t} - T_t(t^{n/2})\right), \phi_{T,r}^{n/2}\right|. \qquad (2.43)
\end{aligned}
$$

Adding and subtracting terms

$$
b_2\left(\mathbf{u}(t^{n/2}) + \mathscr{X}(\mathbf{u}_r^n) + \mathscr{X}(\mathbf{u}(t^n)), \frac{T(t^{n+1}) + T(t^n)}{2}, \phi_{T,r}^{n/2}\right)
$$

to the nonlinear terms in (2.43) leads to

$$
\begin{aligned}
&b_2(\mathbf{u}(t^{n/2}), T(t^{n/2}), \phi_{T,r}^{n/2}) - b_2(\mathscr{X}(\mathbf{u}_r^n), T_r^{n/2}, \phi_{T,r}^{n/2}) \\
&= b_2(\mathscr{X}(\mathbf{u}_r^n), {\eta_T}^{n/2}, \phi_{T,r}^{n/2}) - b_2(\mathscr{X}(\mathbf{u}_r^n), \phi_{T,r}^{n/2}, \phi_{T,r}^{n/2}) \\
&+b_2\left(\mathscr{X}(\mathbf{e}_{\mathbf{u},r}^n), \frac{T(t^{n+1}) + T(t^n)}{2}, \phi_{T,r}^{n/2}\right) \\
&+b_2\left(\mathbf{u}(t^{n/2}), \frac{T(t^{n+1}) + T(t^n)}{2} - T_{n/2}, \phi_{T,r}^{n/2}\right) \\
&+b_2\left(\mathbf{u}(t^{n/2}) - \mathscr{X}(\mathbf{u}(t^n)), \frac{T(t^{n+1}) + T(t^n)}{2}, \phi_{T,r}^{n/2}\right) \\
&+b_2\left(\mathbf{u}(t^{n/2}), T(t^{n/2}) - \frac{T(t^{n+1}) + T(t^n)}{2}, \phi_{T,r}^{n/2}\right).
\end{aligned}
$$

Note that $b_2(\mathbf{u}^{n/2}, \phi_{T,r}^{n+1}, \phi_{T,r}^{n+1}) = 0$. Using Cauchy–Schwarz and Young's inequalities, we obtain

$$
\begin{aligned}
&\frac{1}{2\Delta t}\|\phi_{T,r}^{n+1}\|^2 + \gamma\|\nabla\phi_{T,r}^{n/2}\|^2 + \alpha_2\|(I-P_{T,R})\nabla\phi_{T,r}^{n/2}\|^2 \le \frac{1}{2\Delta t}\|\phi_{T,r}^{n}\|^2 \\
&+\gamma|(\nabla{\eta_T}^{n/2}, \nabla\phi_{T,r}^{n/2})| + \gamma\left|\left(\nabla\left(\frac{T(t^{n+1})+T(t^n)}{\Delta t} - T(t^{n/2})\right), \nabla\phi_{T,r}^{n/2}\right)\right|^2 \\
&+|b_2(\mathscr{X}(\mathbf{u}_r^n), {\eta_T}^{n/2}, \phi_{T,r}^{n/2})| + \left|b_2\left(\mathscr{X}(\mathbf{e}_{\mathbf{u},r}^n), \frac{T(t^{n+1})+T(t^n)}{2}, \phi_{T,r}^{n/2}\right)\right| \\
&+\left|b_2\left(\mathbf{u}(t^{n/2}), \frac{T(t^{n+1})+T(t^n)}{2} - T_{n/2}, \phi_{T,r}^{n/2}\right)\right| \\
&+\left|b_2\left(\mathbf{u}(t^{n/2}) - \mathscr{X}(\mathbf{u}(t^n)), \frac{T(t^{n+1})+T(t^n)}{2}, \phi_{T,r}^{n/2}\right)\right| \\
&+\left|b_2\left(\mathbf{u}(t^{n/2}), T(t^{n/2}) - \frac{T(t^{n+1})+T(t^n)}{2}, \phi_{T,r}^{n/2}\right)\right| \\
&+\alpha_2|\left((I-P_{T,R})\nabla\eta_T^{n/2}, (I-P_{T,R})\nabla\phi_{T,r}^{n/2}\right)| \\
&+\alpha_2|\left((I-P_{T,R})\nabla T(t^{n/2}), (I-P_{T,R})\nabla\phi_{T,r}^{n/2}\right)| \\
&+\alpha_2\left|\left((I-P_{T,R})\nabla\left(\frac{T(t^{n+1})+T(t^n)}{\Delta t} - T(t^{n/2})\right), (I-P_{T,R})\nabla\phi_{T,r}^{n/2}\right)\right| \\
&+|(\frac{T(t^{n+1})-T(t^n)}{\Delta t} - T_t(t^{n/2}), \phi_{T,r}^{n/2})|. \qquad (2.44)
\end{aligned}
$$

Next, we bound the second and third terms on the right-hand side of (2.44), by using Lemma 2.2, Cauchy–Schwarz, Young's and Poincaré's inequalities:

$$
\gamma|(\nabla{\eta_T}^{n/2}, \nabla\phi_{T,r}^{n/2})| \le K\gamma\|\nabla{\eta_T}^{n/2}\|^2 + \frac{\gamma}{6}\|\nabla\phi_{T,r}^{n/2}\|^2, \qquad (2.45)
$$

$$
\begin{aligned}
\gamma\left|\left(\nabla\left(\frac{T(t^{n+1})+T(t^n)}{\Delta t} - T(t^{n/2})\right), \nabla\phi_{T,r}^{n/2}\right)\right|^2 &\le K\gamma\Delta t^4\|\nabla T_{tt}(t^*)\|^2 \\
&+\frac{\gamma}{6}\|\nabla\phi_{T,r}^{n/2}\|^2. \qquad (2.46)
\end{aligned}
$$

The first nonlinear term on right-hand side of (2.44) can be rearranged by adding and subtracting the term $b_2(\mathscr{X}(\mathbf{u}(t^n)), {\eta_T}^{n/2}, \phi_{T,r}^{n/2})$ as

$$
\begin{aligned}
b_2(\mathscr{X}(\mathbf{u}_r^n), {\eta_T}^{n/2}, \phi_{T,r}^{n/2}) \le |b_2(\mathscr{X}(\eta_{\mathbf{u}}^n), {\eta_T}^{n/2}, \phi_{T,r}^{n/2})| + |b_2(\mathscr{X}(\phi_{\mathbf{u},r}), {\eta_T}^{n/2}, \phi_{T,r}^{n/2})| \\
+|b_2(\mathscr{X}(\mathbf{u}(t^n)), {\eta_T}^{n/2}, \phi_{T,r}^{n/2})|. \qquad (2.47)
\end{aligned}
$$

To bound the terms on the right-hand side of (2.47), we use Lemma 2.1 and Young's inequality:

$$|b_2(\mathscr{X}(\eta_{\mathbf{u}}^n), {\eta_T}^{n/2}, \phi_{T,r}^{n/2})| \leq K\gamma^{-1}(\|\mathbb{D}\eta_{\mathbf{u}}^n\|^2 + \|\mathbb{D}\eta_{\mathbf{u}}^{n-1}\|^2)\|\nabla\eta_T^{n/2}\|^2 + \frac{\gamma}{6}\|\nabla\phi_{T,r}^{n/2}\|^2,$$

$$|b_2(\mathscr{X}(\phi_{\mathbf{u},r}), {\eta_T}^{n/2}, \phi_{T,r}^{n/2})| \leq K\gamma^{-1}h^{-1}(\|\phi_{\mathbf{u},r}^n\|^2 + \|\phi_{\mathbf{u},r}^{n-1}\|^2)\|\nabla\eta_T^{n/2}\|^2 + \frac{\gamma}{6}\|\nabla\phi_{T,r}^{n/2}\|^2,$$

$$|b_2(\mathscr{X}(\mathbf{u}(t^n)), {\eta_T}^{n/2}, \phi_{T,r}^{n/2})| \leq K\gamma^{-1}(\|\mathbb{D}\mathbf{u}(t^n)\|^2 + \|\mathbb{D}\mathbf{u}(t^{n-1})\|^2)\|\nabla\eta_T^{n/2}\|^2 + \frac{\gamma}{6}\|\nabla\phi_{T,r}^{n/2}\|^2.$$

Using similar techniques for the other red nonlinear terms on the right-hand side of (2.44), we get

$$\left|b_2(\mathscr{X}(\mathbf{e}_{\mathbf{u}}^n), \frac{T(t^{n+1}) + T(t^n)}{2}, \phi_{T,r}^{n/2})\right|$$
$$\leq K\gamma^{-1}(\|\mathbb{D}\eta_{\mathbf{u}}^n\|^2 + \|\mathbb{D}\eta_{\mathbf{u}}^{n-1}\|^2 + h^{-1}(\|\phi_{\mathbf{u},r}^n\|^2 + \|\phi_{\mathbf{u},r}^{n-1})\|^2)\left\|\nabla\left(\frac{T(t^{n+1}) + T(t^n)}{2}\right)\right\|^2$$
$$+\frac{\gamma}{6}\|\nabla(\phi_{T,r}^{n/2})\|^2,$$

$$\left|b_2\left(\mathbf{u}(t^{n/2}), \frac{T(t^{n+1}) + T(t^n)}{2} - T(t^{n/2}), \phi_{T,r}^{n/2}\right)\right|$$
$$\leq K\gamma^{-1}\Delta t^4\|\mathbb{D}(\mathbf{u}(t^{n/2}))\|^2\|\nabla T_{tt}(t^*)\|^2 + \frac{\gamma}{6}\|\nabla\phi_{T,r}^{n/2}\|^2,$$

$$\left|b_2\left(\mathscr{X}(\mathbf{u}(t^n)) - \mathbf{u}(t^{n/2}), \frac{T(t^{n+1}) + T(t^n)}{2}, \phi_{T,r}^{n/2}\right)\right|$$
$$\leq K\gamma^{-1}\|\nabla(\mathscr{X}(\mathbf{u}(t^n)) - \mathbf{u}(t^{n/2}))\|^2\left\|\nabla\left(\frac{T(t^{n+1}) + T(t^n)}{2}\right)\right\|^2 + \frac{\gamma}{6}\|\nabla\phi_{T,r}^{n/2}\|^2,$$

$$\left|b_2\left(\mathbf{u}(t^{n/2}), \frac{T(t^{n+1}) + T(t^n)}{2} - T(t^{n/2}), \phi_{T,r}^{n/2}\right)\right|$$
$$\leq K\gamma^{-1}\Delta t^4\|\nabla\mathbf{u}(t^{n/2})\|^2\|\nabla T_{tt}(t^*)\|^2 + \frac{\gamma}{6}\|\nabla\phi_{T,r}^{n/2}\|^2. \tag{2.48}$$

The ninth, tenth and eleventh terms of (2.44) can be bounded by using the fact that $\|(I - P_{T,R})\nabla w\|^2 \leq \|\nabla w\|^2$:

$$
\begin{aligned}
&\alpha_2 \left| \left((I - P_{T,R})\nabla \left(\frac{T(t^{n+1}) + T(t^n)}{\Delta t} - T(t^{n/2}) \right), (I - P_{T,R})\nabla \phi_{T,r}^{n/2} \right) \right| \\
&+\alpha_2 \left| \left((I - P_{T,R})\nabla \eta_T^{n/2}, (I - P_{T,R})\nabla \phi_{T,r}^{n/2} \right) \right| \\
&+\alpha_2 \left| \left((I - P_{T,R})\nabla T^{n/2}, (I - P_{T,R})\nabla \phi_{T,r}^{n/2} \right) \right| \\
&\leq \alpha_2 \|\nabla \eta_T^{n/2}\|^2 + \alpha_2 \|(I - P_{T,R})\nabla T^{n/2}\|^2 + \alpha_2 \Delta t^4 \|(I - P_{T,R})\nabla T_{tt}\|^2 \\
&+\frac{\alpha_2}{2} \|(I - P_{T,R})\nabla \phi_{T,r}^{n/2}\|^2 . \qquad (2.49)
\end{aligned}
$$

For the last term on the right-hand side of (2.44), we apply Cauchy–Schwarz, Poincaré's, Young's inequalities and Lemma 2.2 as

$$
\left(\frac{T(t^{n+1}) - T(t^n)}{\Delta t} - T_t(t^{n/2}), \phi_{T,r}^{n/2} \right) \leq K\gamma^{-1} \Delta t^4 \|T_{ttt}(t^*)\|^2 + \frac{\gamma}{6} \|\nabla \phi_{T,r}^{n/2}\|^2. \qquad (2.50)
$$

Inserting (2.45)–(2.50) in (2.44), multiplying by $2\Delta t$ and summing over the time steps produces

$$
\begin{aligned}
&\|\phi_{T,r}^M\|^2 + \gamma \Delta t \sum_{n=0}^{M-1} \|\nabla \phi_{T,r}^{n/2}\|^2 + \alpha_2 \Delta t \sum_{n=0}^{M-1} \|(I - P_{T,R})\nabla \phi_{T,r}^{n/2}\|^2 \leq \|\phi_{T,r}^0\|^2 \\
&+K\Delta t \Bigg((\gamma + \alpha_2) \sum_{n=0}^{M-1} \|\nabla \eta_T{}^{n/2}\|^2 + \gamma^{-1} h^{-1} \sum_{n=0}^{M-1} (\|\phi_{\mathbf{u},r}^n\|^2 + \|\phi_{\mathbf{u},r}^{n-1}\|^2) \|\nabla \eta_T^{n/2}\|^2 \\
&+\gamma^{-1} \sum_{n=0}^{M-1} (1 + \|\mathbb{D}\eta_{\mathbf{u}}^n\|^2 + \|\mathbb{D}\eta_{\mathbf{u}}^{n-1}\|^2) \|\nabla \eta_T^{n/2}\|^2 + \gamma^{-1} \Big(\|\mathbb{D}\eta_{\mathbf{u}}^n\|^2 + \|\mathbb{D}\eta_{\mathbf{u}}^{n-1}\|^2 \\
&+h^{-1} (\|\phi_{\mathbf{u},r}^n\|^2 + \|\phi_{\mathbf{u},r}^{n-1}) \|^2 \Big) + \Delta t^4 \Big((\gamma + \gamma^{-1} \|\mathbb{D}(\mathbf{u}(t^{n/2}))\|^2) \|\nabla T_{tt}(t^*)\|^2 \\
&+\gamma^{-1} \|T_{ttt}(t^*)\|^2 + \alpha_2 \|(I - P_{T,R})\nabla T_{tt}(t^*)\|^2 \Big) + \alpha_2 \|(I - P_{T,R})\nabla T(t^{n/2})\|^2 \Bigg). \qquad (2.51)
\end{aligned}
$$

Using Lemma 2.3, Lemma 2.4, Assumption 1 and regularity assumptions in (2.51) results in

$$
\begin{aligned}
&\|\phi_{T,r}^M\|^2 + \gamma \Delta t \sum_{n=0}^{M-1} \|\nabla \phi_{T,r}^{n/2}\|^2 + \alpha_2 \Delta t \sum_{n=0}^{M-1} \|(I - P_{T,R})\nabla \phi_{T,r}^{n/2}\|^2 \\
&\leq \|\phi_{T,r}^0\|^2 + K \Big(h^{2m} + (\|S_{T,r}\|_2 + \|S_{T,R}\|_2) h^{2m+2} + \varepsilon_{T,r}^2 + \varepsilon_{T,R}^2
\end{aligned}
$$

$$+(1 + h^{2m} + \|S_{\mathbf{u},r}\|_2 h^{2m+2} + \varepsilon_{\mathbf{u},r}^2)(h^{2m} + \|S_{T,r}\|_2 h^{2m+2} + \varepsilon_{T,r}^2)$$

$$+(\Delta t)^4 + \gamma^{-1}h^{-1}\sum_{n=0}^{M-1}(\|\phi_{\mathbf{u},r}^n\|^2 + \|\phi_{\mathbf{u},r}^{n-1}\|^2)\|\nabla\eta_T^{n/2}\|^2\Big). \tag{2.52}$$

Similarly, the error estimation for the concentration is given by

$$\|\phi_{C,r}^M\|^2 + D_c\Delta t\sum_{n=0}^{M-1}\|\nabla\phi_{C,r}^{n/2}\|^2 + \alpha_3\Delta t\sum_{n=0}^{M-1}\|(I - P_{C,R})\nabla\phi_{C,r}^{n/2}\|^2$$

$$\leq \|\phi_{C,r}^0\|^2 + K\Big(h^{2m} + (\|S_{C,r}\|_2 + \|S_{C,R}\|_2)h^{2m+2} + \varepsilon_{C,r}^2 + \varepsilon_{C,R}^2$$

$$+(1 + h^{2m} + \|S_{\mathbf{u},r}\|_2 h^{2m+2} + \varepsilon_{\mathbf{u},r}^2)(h^{2m} + \|S_{C,r}\|_2 h^{2m+2} + \varepsilon_{C,r}^2)$$

$$+(\Delta t)^4 + D_c^{-1}h^{-1}\sum_{n=0}^{M-1}(\|\phi_{\mathbf{u},r}^n\|^2 + \|\phi_{\mathbf{u},r}^{n-1}\|^2)\|\nabla\eta_C^{n/2}\|^2\Big). \tag{2.53}$$

To obtain an estimation for the velocity we use similar arguments as above. Thus, in a similar manner, for the velocity, we add and subtract terms:

$$2\nu\left(\mathbb{D}\left(\frac{\mathbf{u}(t^{n+1}) + \mathbf{u}(t^n)}{2}\right), \mathbb{D}\mathbf{v}_r\right) + \left(Da^{-1}\left(\frac{\mathbf{u}(t^{n+1}) + \mathbf{u}(t^n)}{2}\right), \mathbf{v}_r\right)$$

$$-\left(\frac{p(t^{n+1}) + p(t^n)}{2}, \nabla\cdot\mathbf{v}_r\right)$$

$$+\alpha_1\left((I - P_{\mathbf{u},R})\mathbb{D}\left(\frac{\mathbf{u}(t^{n+1}) + \mathbf{u}(t^n)}{2}\right), (I - P_{\mathbf{u},R})\mathbb{D}\mathbf{v}_r\right)$$

$$-\beta_T\left(\mathbf{g}\left(\frac{T(t^{n+1}) + T(t^n)}{2}\right), \mathbf{v}_r\right) - \beta_C\left(\mathbf{g}\left(\frac{C(t^{n+1}) + C(t^n)}{2}\right), \mathbf{v}_r\right)$$

$$b_1\left(\mathbf{u}(t^{n/2}) + \mathscr{X}(\mathbf{u}_r{}^n) + \mathscr{X}(\mathbf{u}(t^n)), \frac{\mathbf{u}(t^{n+1}) + \mathbf{u}(t^n)}{2}, \mathbf{v}_r\right)$$

to (2.36). Letting $\mathbf{v}_r = \boldsymbol{\phi}_{\mathbf{u},r}^{n/2}$ in (2.25), and applying the polarization identity gives

$$\frac{1}{2\Delta t}\|\boldsymbol{\phi}_{\mathbf{u},r}^{n+1}\|^2 - \frac{1}{2\Delta t}\|\boldsymbol{\phi}_{\mathbf{u},r}^n\|^2 + \frac{1}{2\Delta t}\|\boldsymbol{\phi}_{\mathbf{u},r}^{n+1} - \boldsymbol{\phi}_{\mathbf{u},r}^n\|^2 + 2\nu\|\mathbb{D}\boldsymbol{\phi}_{\mathbf{u},r}^{n/2}\|^2$$

$$+\alpha_1\|(I - P_{\mathbf{u},R})\mathbb{D}\boldsymbol{\phi}_{\mathbf{u},r}^{n/2}\|^2 + Da^{-1}\|\boldsymbol{\phi}_{\mathbf{u},r}^{n/2}\|^2$$

$$\leq |\beta_T(\mathbf{g}(\eta_T^{n/2}), \boldsymbol{\phi}_{\mathbf{u},r}^{n/2})| + |\beta_C(\mathbf{g}(\eta_C^{n/2}), \boldsymbol{\phi}_{\mathbf{u},r}^{n/2})| + |\beta_T(\mathbf{g}(\phi_{T,r}^{n/2}), \boldsymbol{\phi}_{\mathbf{u},r}^{n/2})|$$

$$+|\beta_C(\mathbf{g}(\phi_{C,r}^{n/2}), \boldsymbol{\phi}_{\mathbf{u},r}^{n/2})| + \left|\beta_T\left(\mathbf{g}\left(\frac{T(t^{n+1}) + T(t^n)}{2} - T^{n/2}\right), \boldsymbol{\phi}_{\mathbf{u},r}^{n/2}\right)\right|$$

$$
\begin{aligned}
&+\left|\beta_C\left(\mathbf{g}\left(\frac{C(t^{n+1})+C(t^n)}{2}-C^{n/2}\right),\boldsymbol{\phi}_{\mathbf{u},r}^{n/2}\right)\right|+\left|\left(\frac{\boldsymbol{\eta}_u^{n+1}-\boldsymbol{\eta}_u^n}{\Delta t},\boldsymbol{\phi}_{\mathbf{u},r}^{n/2}\right)\right|\\
&+2\nu|(\mathbb{D}\boldsymbol{\eta}_{\mathbf{u}}{}^{n/2},\mathbb{D}\boldsymbol{\phi}_{\mathbf{u},r}^{n/2})|+2\nu\left|\left(\mathbb{D}\left(\frac{\mathbf{u}(t^{n+1})+\mathbf{u}(t^n)}{2}\right)-\mathbb{D}\mathbf{u}(t^{n/2}),\mathbb{D}\boldsymbol{\phi}_{\mathbf{u},r}^{n/2}\right)\right|\\
&+Da^{-1}|(\boldsymbol{\eta}_{\mathbf{u}}{}^{n/2},\boldsymbol{\phi}_{\mathbf{u},r}^{n/2})|+Da^{-1}\left|\left(\left(\frac{\mathbf{u}(t^{n+1})+\mathbf{u}(t^n)}{2}\right)-\mathbf{u}(t^{n/2}),\boldsymbol{\phi}_{\mathbf{u},r}^{n/2}\right)\right|\\
&+\alpha_1\left|\left((I-P_{\mathbf{u},R})\nabla\eta_{\mathbf{u}}^{n/2},(I-P_{\mathbf{u},R})\mathbb{D}\phi_{\mathbf{u},r}^{n/2}\right)\right|\\
&+\alpha_1\left|\left((I-P_{\mathbf{u},R})\mathbb{D}\mathbf{u}(t^{n/2}),(I-P_{\mathbf{u},R})\mathbb{D}\boldsymbol{\phi}_{\mathbf{u},r}^{n/2}\right)\right|\\
&+\alpha_1\left|\left((I-P_{\mathbf{u},R})\mathbb{D}\left(\frac{\mathbf{u}(t^{n+1})+\mathbf{u}(t^n)}{2}-\mathbf{u}(t^{n/2})\right),(I-P_{\mathbf{u},R})\mathbb{D}\boldsymbol{\phi}_{\mathbf{u},r}^{n/2}\right)\right|\\
&+|b_1(\mathscr{X}(\mathbf{u}_r^n),\eta_{\mathbf{u}}{}^{n/2},\phi_{\mathbf{u},r}^{n/2})|+\left|b_1\left(\mathscr{X}(\eta_{\mathbf{u}}^n),\frac{\mathbf{u}(t^{n+1})+\mathbf{u}(t^n)}{2},\phi_{\mathbf{u},r}^{n/2}\right)\right|\\
&+\left|b_1\left(\mathscr{X}(\boldsymbol{\phi}_{\mathbf{u},r}^n),\frac{\mathbf{u}(t^{n+1})+\mathbf{u}(t^n)}{2},\phi_{\mathbf{u},r}^{n/2}\right)\right|\\
&+\left|b_1\left(\mathbf{u}(t^{n/2}),\frac{\mathbf{u}(t^{n+1})+\mathbf{u}(t^n)}{2}-\mathbf{u}(t^{n/2}),\phi_{\mathbf{u},r}^{n/2}\right)\right|\\
&+\left|b_1\left((\mathscr{X}(\mathbf{u}(t^n))-\mathbf{u}(t^{n/2}),\frac{\mathbf{u}(t^{n+1})+\mathbf{u}(t^n)}{2},\phi_{\mathbf{u},r}^{n/2}\right)\right|\\
&+\left|\left(\frac{p(t^{n+1})+p(t^n)}{2}-p(t^{n/2}),\nabla\cdot\boldsymbol{\phi}_{\mathbf{u},r}^{n/2}\right)\right|\\
&+\left|\left(\frac{p(t^{n+1})+p(t^n)}{2}-q_h,\nabla\cdot\boldsymbol{\phi}_{\mathbf{u},r}^{n/2}\right)\right|+\left|\left(\frac{\mathbf{u}^{n+1}-\mathbf{u}^n}{\Delta t}-\mathbf{u}_t^{n+1},\boldsymbol{\phi}_{\mathbf{u},r}^{n/2}\right)\right|.
\end{aligned}
\tag{2.54}
$$

Note that $\left(\frac{\boldsymbol{\eta}_u^{n+1}-\boldsymbol{\eta}_u^n}{\Delta t},\boldsymbol{\phi}_{\mathbf{u},r}^{n/2}\right)=0$ due to the definition of the L^2 projection. Each of the terms in (2.54) can be bounded in a similar manner. Thus, one gets

$$
\begin{aligned}
&|\beta_T(\mathbf{g}(\eta_T^{n/2}),\boldsymbol{\phi}_{\mathbf{u},r}^{n/2})|+|\beta_C(\mathbf{g}(\eta_C^{n/2}),\boldsymbol{\phi}_{\mathbf{u},r}^{n/2})|+|\beta_T(\mathbf{g}(\phi_{T,r}^{n/2}),\boldsymbol{\phi}_{\mathbf{u},r}^{n/2})|\\
&\quad+|\beta_C(\mathbf{g}(\phi_{C,r}^{n/2}),\boldsymbol{\phi}_{\mathbf{u},r}^{n/2})|+\left|\beta_T\left(\mathbf{g}\left(\frac{T(t^{n+1})+T(t^n)}{2}-T(t^{n/2})\right),\boldsymbol{\phi}_{\mathbf{u},r}^{n/2}\right)\right|\\
&\quad+|\beta_C(\mathbf{g}(\frac{C(t^{n+1})+C(t^n)}{2}-C(t^{n/2})),\boldsymbol{\phi}_{\mathbf{u},r}^{n/2})|\\
&\leq K\nu^{-1}\|\mathbf{g}\|_\infty^2\Big(\beta_T^2(\|\eta_T^{n/2}\|^2+\|\phi_{T,r}^{n/2}\|^2+\Delta t^4\|T_{tt}(\cdot,\tilde{t})\|^2)\\
&\quad+\beta_C^2(\|\eta_C^{n/2}\|^2+\|\phi_{C,r}^{n/2}\|^2+\Delta t^4\|C_{tt}(t^*)\|^2)\Big)+\frac{\nu}{10}\|\mathbb{D}\boldsymbol{\phi}_{\mathbf{u},r}^{n/2}\|^2,
\end{aligned}
\tag{2.55}
$$

$$2\nu|(\mathbb{D}\boldsymbol{\eta}_{\mathbf{u}}^{n/2}, \mathbb{D}\boldsymbol{\phi}_{\mathbf{u},r}^{n/2})| + 2\nu\Big|\Big(\mathbb{D}\Big(\frac{\mathbf{u}(t^{n+1}) + \mathbf{u}(t^n)}{2}\Big) - \mathbb{D}\mathbf{u}(t^{n/2}), \mathbb{D}\boldsymbol{\phi}_{\mathbf{u},r}^{n/2}\Big)\Big|$$
$$\leq K\nu\Big(\|\mathbb{D}\boldsymbol{\eta}_{\mathbf{u}}^{n/2}\|^2 + \Delta t^4\|\mathbb{D}\mathbf{u}_{tt}(\cdot,\tilde{t})\|^2\Big) + \frac{\nu}{10}\|\mathbb{D}\boldsymbol{\phi}_{\mathbf{u},r}^{n/2}\|^2, \tag{2.56}$$

$$Da^{-1}|(\boldsymbol{\eta}_{\mathbf{u}}^{n/2}, \boldsymbol{\phi}_{\mathbf{u},r}^{n/2})| + Da^{-1}|\Big((\frac{\mathbf{u}(t^{n+1}) + \mathbf{u}(t^n)}{2}) - \mathbf{u}(t^{n/2}), \boldsymbol{\phi}_{\mathbf{u},r}^{n/2}\Big)|$$
$$\leq K Da^{-1}\Big(\|\boldsymbol{\eta}_{\mathbf{u}}^{n/2}\|^2 + \Delta t^4\|\mathbf{u}_{tt}(t^*)\|^2\Big) + \frac{Da^{-1}}{2}\|\boldsymbol{\phi}_{\mathbf{u},r}^{n/2}\|^2, \tag{2.57}$$

$$|(\frac{p(t^{n+1}) + p(t^n)}{2} - p(t^{n/2}), \nabla\cdot\boldsymbol{\phi}_{\mathbf{u},r}^{n/2})| + |(\frac{p(t^{n+1}) + p(t^n)}{2} - q_h, \nabla\cdot\boldsymbol{\phi}_{\mathbf{u},r}^{n/2})|$$
$$\leq K\nu^{-1}(\Delta t^4\left\|p_{tt}(t^*)\right\|^2 + \left\|\frac{p(t^{n+1}) + p(t^n)}{2} - q_h\right\|^2) + \frac{\nu}{10}\left\|\mathbb{D}\boldsymbol{\phi}_{\mathbf{u},r}^{n/2}\right\|^2, \tag{2.58}$$

$$|(\mathbf{u}_t^{n+1} - \frac{\mathbf{u}^{n+1} - \mathbf{u}^n}{\Delta t}, \boldsymbol{\phi}_{\mathbf{u},r}^{n/2})| \leq K\nu^{-1}\Delta t^4\|\mathbf{u}_{ttt}(t^*)\|^2 + \frac{\nu}{10}\left\|\mathbb{D}\boldsymbol{\phi}_{\mathbf{u},r}^{n/2}\right\|^2, \tag{2.59}$$

$$\alpha_1|\Big((I - P_{\mathbf{u},R})\mathbb{D}(\frac{\mathbf{u}(t^{n+1}) + \mathbf{u}(t^n)}{2} - \mathbf{u}(t^{n/2})), (I - P_{\mathbf{u},R})\mathbb{D}\boldsymbol{\phi}_{\mathbf{u},r}^{n/2}\Big)|$$
$$+\alpha_1|\Big((I - P_{\mathbf{u},R})\nabla\eta_{\mathbf{u}}^{n/2}, (I - P_{\mathbf{u},R})\mathbb{D}\phi_{\mathbf{u},r}^{n/2}\Big)|$$
$$+\alpha_1|\Big((I - P_{\mathbf{u},R})\mathbb{D}\mathbf{u}(t^{n/2}), (I - P_{\mathbf{u},R})\mathbb{D}\boldsymbol{\phi}_{\mathbf{u},r}^{n/2}\Big)|$$
$$\leq K\alpha_1\Big(\Delta t^4\|(I - P_{\mathbf{u},R})\mathbb{D}\mathbf{u}_{tt}(t^*)\|^2 + \|\nabla\eta_{\mathbf{u}}^{n/2}\|^2 + \|(I - P_{\mathbf{u},R})\mathbb{D}\mathbf{u}(t^{n/2})\|^2\Big)$$
$$+\frac{\alpha_1}{2}\|(I - P_{\mathbf{u},R})\mathbb{D}\phi_{\mathbf{u},r}^{n/2}\|^2. \tag{2.60}$$

The first nonlinear term in (2.61) is organized as

$$|b_1(\mathscr{X}(\mathbf{u}_r^n), {\eta_{\mathbf{u}}}^{n/2}, \phi_{\mathbf{u},r}^{n/2})| \leq |b_1(\mathscr{X}(\mathbf{u}(t^n)), {\eta_{\mathbf{u}}}^{n/2}, \phi_{\mathbf{u},r}^{n/2})|$$
$$+|b_1(\mathscr{X}(\eta_{\mathbf{u}}^n), {\eta_{\mathbf{u}}}^{n/2}, \phi_{\mathbf{u},r}^{n/2})| + |b_1(\mathscr{X}(\phi_{\mathbf{u},r}^n), {\eta_{\mathbf{u}}}^{n/2}, \phi_{\mathbf{u},r}^{n/2})|. \tag{2.61}$$

The terms on the right-hand side of (2.61) are bounded as before:

$$|b_1(\mathscr{X}(\mathbf{u}(t^n)), {\eta_{\mathbf{u}}}^{n/2}, \phi_{\mathbf{u},r}^{n/2})|$$
$$\leq K\nu^{-1}(\|\mathbb{D}\mathbf{u}(t^n)\|^2 + \|\mathbb{D}\mathbf{u}(t^{n-1})\|^2)\|\mathbb{D}{\eta_{\mathbf{u}}}^{n/2}\|^2 + \frac{\nu}{10}\|\mathbb{D}\boldsymbol{\phi}_{\mathbf{u},r}^{n/2}\|^2,$$
$$|b_1(\mathscr{X}(\eta_{\mathbf{u}}^n), {\eta_{\mathbf{u}}}^{n/2}, \phi_{\mathbf{u},r}^{n/2})|$$

$$\leq K\nu^{-1}(\|\mathbb{D}\eta_{\mathbf{u}}^{n}\|^2+\|\mathbb{D}\eta_{\mathbf{u}}^{n-1}\|^2)\|\mathbb{D}\eta_{\mathbf{u}}{}^{n/2}\|^2+\frac{\nu}{10}\|\mathbb{D}\boldsymbol{\phi}_{\mathbf{u},r}^{n/2}\|^2,$$

$$|b_1(\mathscr{X}(\phi_{\mathbf{u},r}^{n}),\eta_{\mathbf{u}}{}^{n/2},\phi_{\mathbf{u},r}^{n/2})|$$

$$\leq K\nu^{-1}h^{-1}(\|\phi_{\mathbf{u},r}^{n}\|^2+\|\phi_{\mathbf{u},r}^{n-1})\|^2)\|\mathbb{D}\eta_{\mathbf{u}}{}^{n/2}\|^2+\frac{\nu}{10}\|\mathbb{D}\boldsymbol{\phi}_{\mathbf{u},r}^{n/2}\|^2.$$

Similarly, the remaining nonlinear terms can be bounded as

$$\Big|b_1\Big(\Big(\mathscr{X}(\boldsymbol{\eta}_{\mathbf{u}}^{n}),\frac{\mathbf{u}(t^{n+1})+\mathbf{u}(t^{n})}{2},\phi_{\mathbf{u},r}^{n/2}\Big)\Big|+\Big|b_1\Big(\Big(\mathscr{X}(\boldsymbol{\phi}_{\mathbf{u},r}^{n}),\frac{\mathbf{u}(t^{n+1})+\mathbf{u}(t^{n})}{2},\phi_{\mathbf{u},r}^{n/2}\Big)\Big|$$

$$\leq K\nu^{-1}\Big(\|\mathbb{D}\boldsymbol{\eta}_{\mathbf{u}}^{n}\|^2+\|\mathbb{D}\boldsymbol{\eta}_{\mathbf{u}}^{n-1}\|^2+h^{-1}(\|\boldsymbol{\phi}_{\mathbf{u},r}^{n}\|^2+\|\boldsymbol{\phi}_{\mathbf{u},r}^{n-1}\|^2)\Big)\Big\|\mathbb{D}\Big(\frac{\mathbf{u}(t^{n+1})+\mathbf{u}(t^{n})}{2}\Big)\Big\|^2$$

$$+\frac{\nu}{10}\|\mathbb{D}\boldsymbol{\phi}_{\mathbf{u},r}^{n/2}\|^2,$$

$$|b_1(\mathbf{u}(t^{n/2}),\frac{\mathbf{u}(t^{n+1})+\mathbf{u}(t^{n})}{2}-\mathbf{u}(t^{n/2}),\phi_{\mathbf{u},r}^{n/2})|$$

$$\leq K\nu^{-1}\Delta t^4\|\mathbb{D}\mathbf{u}(t^{n/2})\|^2\|\mathbb{D}\mathbf{u}_{tt}(t^*)\|^2+\frac{\nu}{10}\|\mathbb{D}\boldsymbol{\phi}_{\mathbf{u},r}^{n/2}\|^2,$$

$$b_1((\mathscr{X}(\mathbf{u}(t^{n}))-\mathbf{u}(t^{n/2}),\frac{\mathbf{u}(t^{n+1})+\mathbf{u}(t^{n})}{2},\phi_{\mathbf{u},r}^{n/2})|$$

$$\leq K\nu^{-1}\|\mathbb{D}(\mathscr{X}(\mathbf{u}(t^{n}))-\mathbf{u}(t^{n/2}))\|^2\Big\|\mathbb{D}\Big(\frac{\mathbf{u}(t^{n+1})+\mathbf{u}(t^{n})}{2}\Big)\Big\|^2+\frac{\nu}{10}\|\mathbb{D}\boldsymbol{\phi}_{\mathbf{u},r}^{n/2}\|^2. \quad (2.62)$$

We now insert (2.55)–(2.62) into (2.54) and use regularity assumptions to get

$$\frac{1}{2\Delta t}\|\boldsymbol{\phi}_{\mathbf{u},r}^{n+1}\|^2-\frac{1}{2\Delta t}\|\boldsymbol{\phi}_{\mathbf{u},r}^{n}\|^2+\frac{1}{2\Delta t}\|\boldsymbol{\phi}_{\mathbf{u},r}^{n+1}-\boldsymbol{\phi}_{\mathbf{u},r}^{n}\|^2$$

$$+\nu\|\mathbb{D}\boldsymbol{\phi}_{\mathbf{u},r}^{n/2}\|^2+\frac{Da^{-1}}{2}\|\boldsymbol{\phi}_{\mathbf{u},r}^{n/2}\|^2+\frac{\alpha_1}{2}\|(I-P_{\mathbf{u},R})\mathbb{D}\phi_{\mathbf{u},r}^{n/2}\|^2$$

$$\leq K\Big(\nu^{-1}\|\mathbf{g}\|_\infty^2(\beta_T^2\|\phi_{T,r}^{n/2}\|^2+\beta_C^2\|\phi_{C,r}^{n/2}\|^2)+\|\eta_T^{n/2}\|^2+\|\eta_C^{n/2}\|^2$$

$$+\|\boldsymbol{\eta}_{\mathbf{u}}{}^{n/2}\|^2+\|\mathbb{D}\boldsymbol{\eta}_{\mathbf{u}}{}^{n/2}\|^2(1+\|\mathbb{D}\boldsymbol{\eta}_{\mathbf{u}}^{n}\|^2+\|\mathbb{D}\boldsymbol{\eta}_{\mathbf{u}}^{n-1}\|^2)$$

$$+\nu^{-1}h^{-1}\Big(\|\mathbb{D}\boldsymbol{\eta}_{\mathbf{u}}{}^{n/2}\|^2+\Big\|\mathbb{D}\Big(\frac{\mathbf{u}(t^{n+1})+\mathbf{u}(t^{n})}{2}\Big)\Big\|^2)(\|\boldsymbol{\phi}_{\mathbf{u},r}^{n}\|^2+\|\boldsymbol{\phi}_{\mathbf{u},r}^{n-1}\|^2)$$

$$+\alpha_1\|(I-P_{\mathbf{u},R})\mathbb{D}\mathbf{u}(t^{n/2})\|^2+\|\frac{p(t^{n+1})+p(t^{n})}{2}-q_h\|^2$$

$$+\nu^{-1}\|\mathbf{g}\|_\infty^2\Delta t^4\Big(\beta_T^2\|T_{tt}(\cdot,\tilde{t})\|^2+\beta_C^2\|C_{tt}(t^*)\|^2\Big)$$

$$+(\nu+\nu^{-1}\|\mathbb{D}\mathbf{u}(t^{n/2})\|^2)\Delta t^4\|\mathbb{D}\mathbf{u}_{tt}(\cdot,\tilde{t})\|^2+Da^{-1}\Delta t^4\|\mathbf{u}_{tt}(t^*)\|^2$$

$$+\nu^{-1}\Delta t^4 \left\| p_{tt}(t^*) \right\|^2 + \nu^{-1}\Delta t^4 \|\mathbf{u}_{ttt}(t^*)\|^2$$

$$+\Delta t^4 \alpha_1 \|(I - P_{\mathbf{u},R})\mathbb{D}\mathbf{u}_{tt}(t^*)\|^2\Big). \tag{2.63}$$

Dropping the third term on the left-hand side of (2.63) and summing over the time steps and multiplying by $2\Delta t$ gives

$$\|\boldsymbol{\phi}_{\mathbf{u},r}^{M}\|^2 + \Delta t \sum_{n=0}^{M-1} \Big(2\nu \|\mathbb{D}\boldsymbol{\phi}_{\mathbf{u},r}^{n/2}\|^2 + Da^{-1}\|\boldsymbol{\phi}_{\mathbf{u},r}^{n/2}\|^2 + \alpha_1 \|(I - P_{\mathbf{u},R})\mathbb{D}\boldsymbol{\phi}_{\mathbf{u},r}^{n/2}\|^2\Big)$$

$$\leq \|\boldsymbol{\phi}_{\mathbf{u},r}^{0}\|^2 + K\Delta t \Bigg(\nu^{-1}\|\mathbf{g}\|_{\infty}^2 \sum_{n=0}^{M-1} (\beta_T^2 \|\phi_{T,r}^{n/2}\|^2 + \beta_C^2 \|\phi_{C,r}^{n/2}\|^2)$$

$$+ \sum_{n=0}^{M-1} (\|\eta_T^{n/2}\|^2 + \|\eta_C^{n/2}\|^2 + \|\boldsymbol{\eta}_{\mathbf{u}}{}^{n/2}\|^2)$$

$$+ \sum_{n=0}^{M-1} \|\mathbb{D}\boldsymbol{\eta}_{\mathbf{u}}{}^{n/2}\|^2 (1 + \|\mathbb{D}\boldsymbol{\eta}_{\mathbf{u}}^{n}\|^2 + \|\mathbb{D}\boldsymbol{\eta}_{\mathbf{u}}^{n-1}\|^2)$$

$$+\nu^{-1}h^{-1} \sum_{n=0}^{M-1} \left(\left\|\mathbb{D}\boldsymbol{\eta}_{\mathbf{u}}{}^{n/2}\right\|^2 + \left\|\mathbb{D}\Big(\frac{\mathbf{u}(t^{n+1}) + \mathbf{u}(t^n)}{2}\Big)\right\|^2 \right) (\|\boldsymbol{\phi}_{\mathbf{u},r}^{n}\|^2 + \|\boldsymbol{\phi}_{\mathbf{u},r}^{n-1}\|^2)$$

$$+\alpha_1 \sum_{n=0}^{M-1} \|(I - P_{\mathbf{u},R})\mathbb{D}\mathbf{u}(t^{n/2})\|^2 + \sum_{n=0}^{M-1} \left\| \frac{p(t^{n+1}) + p(t^n)}{2} - q_h \right\|^2$$

$$+\Delta t^4 \Big(\nu^{-1}\|\mathbf{g}\|_{\infty}^2 \sum_{n=0}^{M-1} (\beta_T^2 \|T_{tt}(\cdot,\tilde{t})\|^2 + \beta_C^2 \|C_{tt}(t^*)\|^2)$$

$$+ \sum_{n=0}^{M-1} (\nu + \nu^{-1}\|\mathbb{D}\mathbf{u}(t^{n/2})\|^2)\|\mathbb{D}\mathbf{u}_{tt}(\cdot,\tilde{t})\|^2 + Da^{-1} \sum_{n=0}^{M-1} \|\mathbf{u}_{tt}(t^*)\|^2$$

$$+\nu^{-1} \sum_{n=0}^{M-1} \left\| p_{tt}(t^*) \right\|^2 + \nu^{-1} \sum_{n=0}^{M-1} \|\mathbf{u}_{ttt}(t^*)\|^2 + \alpha_1 \sum_{n=0}^{M-1} \|(I - P_{\mathbf{u},R})\mathbb{D}\mathbf{u}_{tt}(t^*)\|^2\Big)\Bigg).$$

Using Lemma 2.3, Lemma 2.4 and Assumption 1 in (2.64) and applying regularity assumptions leads to

$$\|\boldsymbol{\phi}_{\mathbf{u},r}^{M}\|^2 + \Delta t \sum_{n=0}^{M-1}(2\nu\|\mathbb{D}\boldsymbol{\phi}_{\mathbf{u},r}^{n/2}\|^2 + Da^{-1}\|\boldsymbol{\phi}_{\mathbf{u},r}^{n/2}\|^2)$$

$$+\alpha_1 \sum_{n=0}^{M-1} \|(I - P_{\mathbf{u},R})\mathbb{D}\phi_{\mathbf{u},r}^{n/2}\|^2$$

$$\leq \|\phi_{\mathbf{u},r}^{0}\|^2 + K\Bigg(\nu^{-1}\beta_T^2\|\mathbf{g}\|_\infty^2\Delta t \sum_{n=0}^{M-1} \|\phi_{T,r}^{n/2}\|^2 + \nu^{-1}\beta_C^2\|\mathbf{g}\|_\infty^2\Delta t \sum_{n=0}^{M-1} \|\phi_{C,r}^{n/2}\|^2$$

$$+h^{2m} + (1 + \|S_{\mathbf{u},r}\|_2 + \|S_{\mathbf{u},R}\|_2)h^{2m+2} + \varepsilon_{\mathbf{u},r}^2 + \varepsilon_{\mathbf{u},R}^2 + \sum_{i=r_1+1}^{d} \lambda_i + \sum_{i=r_2+1}^{d} \mu_i$$

$$+ \sum_{i=r_3+1}^{d} \xi_i + (\Delta t)^4 + (h^{2m} + (\|S_{\mathbf{u},r}\|_2 + \|S_{\mathbf{u},R}\|_2)h^{2m+2} + \varepsilon_{\mathbf{u},r}^2)^2$$

$$+\nu^{-1}h^{-1}\Big(h^{2m} + \|S_{\mathbf{u},r}\|h^{2m+2} + \varepsilon_{\mathbf{u},r} + \|\mathbb{D}\mathbf{u}\|_\infty^2\Big) \sum_{n=0}^{M-1}(\|\boldsymbol{\phi}_{\mathbf{u},r}^{n}\|^2)\Bigg). \tag{2.64}$$

Finally, we add (2.52), (2.53) and (2.64) to get

$$\|\boldsymbol{\phi}_{\mathbf{u},r}^{M}\|^2 + \|\phi_{T,r}^{M}\|^2 + \|\phi_{C,r}^{M}\|^2 + \sum_{n=0}^{M-1}\Big(2\nu\Delta t\|\mathbb{D}\boldsymbol{\phi}_{\mathbf{u},r}^{n/2}\|^2 + Da^{-1}\Delta t\|\boldsymbol{\phi}_{\mathbf{u},r}^{n/2}\|^2\Big)$$

$$+\gamma\Delta t \sum_{n=0}^{M-1} \|\nabla\phi_{T,r}^{n/2}\|^2 + D_c\Delta t \sum_{n=0}^{M-1} \|\nabla\phi_{C,r}^{n/2}\|^2 + \alpha_1\|(I - P_{\mathbf{u},R})\mathbb{D}\phi_{\mathbf{u},r}^{n/2}\|^2$$

$$+\alpha_2\Delta t \sum_{n=0}^{M-1} \|(I - P_{T,R})\nabla\phi_{T,r}^{n/2}\|^2 + \alpha_3\Delta t \sum_{n=0}^{M-1} \|(I - P_{C,R})\nabla\phi_{C,r}^{n/2}\|^2$$

$$\leq \|\mathbf{u}_r^0 - \tilde{\mathbf{u}}^0\|^2 + \|T_r^0 - \tilde{T}^0\|^2 + \|C_r^0 - \tilde{C}^0\|^2 + K\Bigg(\nu^{-1}\beta_T^2\|\mathbf{g}\|_\infty^2\Delta t \sum_{n=0}^{M-1} \|\phi_{T,r}^{n/2}\|^2$$

$$+\nu^{-1}\beta_C^2\|\mathbf{g}\|_\infty^2\Delta t \sum_{n=0}^{M-1} \|\phi_{C,r}^{n+1}\|^2+h^{2m}+(\|S_{\mathbf{u},r}\|_2+\|S_{\mathbf{u},R}\|_2+\|S_{T,r}\|_2+\|S_{T,R}\|_2$$

$$+\|S_{C,r}\|_2+\|S_{C,R}\|_2)h^{2m+2}+\varepsilon_{\mathbf{u},r}^2 + \varepsilon_{\mathbf{u},R}^2+\varepsilon_{T,r}^2+\varepsilon_{T,R}^2+\varepsilon_{C,r}^2+\varepsilon_{C,R}^2 + (\Delta t)^4$$

$$+(h^{2m}+\|S_{\mathbf{u},r}\|_2h^{2m+2}+\varepsilon_{\mathbf{u},r}^2) \times \big(h^{2m}+(\|S_{\mathbf{u},r}\|_2+\|S_{T,r}\|_2 + \|S_{C,r}\|_2)h^{2m+2}$$

$$
+\varepsilon_{\mathbf{u},r}^2+\varepsilon_{T,r}^2+\varepsilon_{C,r}^2)+\sum_{i=r_1+1}^{d}\lambda_i+\sum_{i=r_2+1}^{d}\mu_i+\sum_{i=r_3+1}^{d}\xi_i
$$

$$
+\Big((\nu^{-1}+\gamma^{-1}+D_c^{-1})h^{2m-1}+(\|S_{\mathbf{u},r}\|+\|S_{T,r}\|+\|S_{C,r}\|)h^{2m+1}
$$

$$
+\nu^{-1}h^{-1}\varepsilon_{\mathbf{u},r}+\gamma^{-1}h^{-1}\varepsilon_{T,r}+D_c^{-1}h^{-1}\varepsilon_{C,r}+\|\mathbb{D}\mathbf{u}\|_\infty^2\Big)\sum_{n=0}^{M-1}\|\boldsymbol{\phi}_{\mathbf{u},r}^n\|^2\Big).
$$

We remark that the application of the discrete Gronwall inequality requires an assumption on the time step size. The final error estimation can be obtained by using the assumption $(\mathbf{u}_r^0, T_r^0, C_r^0) = (\tilde{\mathbf{u}}^0, \tilde{T}^0, \tilde{C}^0)$, the triangle inequality and (2.21)–(2.22).

2.5 Numerical Experiments

This section presents the numerical experiments for the linearly extrapolated schemes described by (2.25)–(2.27). First of all, we aim to illustrate theoretical error estimate (2.35) numerically and then show that our solution is more accurate than the POD solution.

2.5.1 Problem Description

In our numerical tests, we use the mathematical model given in [4, 8, 18]. The following boundary conditions are used in numerical experiments:

$$
\mathbf{u} = \mathbf{0} \text{ on } \partial\Omega,
$$

$$
T = 0,\ C = 0 \text{ for } x = 0,
$$

$$
T = 1,\ C = 1 \text{ for } x = 1,
$$

$$
\nabla T \cdot n = 0,\ \nabla C \cdot n = 0 \text{ for } y = 0,\ y = 2.
$$

The parameters are chosen as $\Delta t = 1.5625e-05$, $r = 12$, $N = 0.8$, $Le = 2$, $\nu = 1$, $Pr = 1$, $\beta_T = -\frac{Ra}{Pr}$, $\beta_C = \frac{Ra \cdot N}{Pr}$, $\mathbf{g} = \mathbf{e}_2$, $\gamma = \frac{1}{Pr}$, $D_c = \frac{1}{Le \cdot Pr}$. Time domain is $[0, 1]$ and the initial conditions are taken zero. In the computations of the snapshots, we use finite element spatial discretization, Crank Nicholson temporal discretization with $Ra = 10^4$ and $\Delta t = 0.00025$ with 30×60 uniform triangulation. Thus, the total degrees of freedom is 59,255 for Taylor-Hood elements and piecewise quadratics for both temperature and concentration.

2.5.2 Test 1: Convergence Rates with Respect to Δt and R

To measure the efficiency of the method, we compute the rates of convergence with respect to error sources. First, we illustrate error estimate (2.35) in Table 2.1 by scaling the error with respect to Δt. For this test, we use

$$R_u = 18,\ R_T = R_C = 4,\ (\nu_T)_{vel} = 2,\ (\nu_T)_{temp} = (\nu_T)_{conc} = 1,$$

where R_u, R_T, R_C and $(\nu_T)_{vel}$, $(\nu_T)_{temp}$, $(\nu_T)_{conc}$ denote VMS modes numbers and artificial viscosities for the velocity, the temperature and the concentration. The rates in Table 2.1 are approximately 2 which is consistent with the Crank Nicholson time discretization method. We also scale the error with respect to VMS cutoff R. We fix the artificial viscosities as $\alpha_1 = 2$, $\alpha_2 = \frac{1}{8}$, $\alpha_3 = \frac{1}{8}$. The VMS contribution for each variable is given by

$$\varepsilon_{\mathbf{u},R} = \sqrt{\sum_{j=R+1}^{d} \|\psi_j\|_1^2 \lambda_j}, \quad \varepsilon_{T,R} = \sqrt{\sum_{j=R+1}^{d} \|\phi_j\|_1^2 \mu_j}, \quad \varepsilon_{C,R} = \sqrt{\sum_{j=R+1}^{d} \|\eta_j\|_1^2 \xi_j}.$$

The results of this test are given in Table 2.2. This table shows that the rate of convergence of $\|\mathbf{u}-\mathbf{u}_r\|_{L^2(H^1)}$, $\|T-T_r\|_{L^2(H^1)}$, $\|C-C_r\|_{L^2(H^1)}$ with respect to R is close to the theoretical value of 0.5 predicted by (2.35).

Table 2.1 Convergence of the VMS-POD for varying Δt

r	R	Δt	$\|T-T_r\|_{L^2(H^1)}$	Rate	$\|C-C_r\|_{L^2(H^1)}$	Rate
20	4	0.002	38.2096	–	68.4042	–
20	4	0.0005	4.9725	1.47	9.4995	1.42
20	4	0.000125	0.1680	2.44	0.2502	2.62
r	R	Δt	$\|\mathbf{u}-\mathbf{u}_r\|_{L^2(H^1)}$	Rate		
20	18	0.001	1.9257e3	–		
20	18	0.0005	1.8132e2	3.40		
20	18	0.000125	4.1218	2.72		

Table 2.2 Convergence of the VMS-POD for varying R

r	R	ε_u	$\|\mathbf{u}-\mathbf{u}_r\|_{L^2(H^1)}$	Rate	ε_T	$\|T-T_r\|_{L^2(H^1)}$	Rate
12	4	21.8237	0.63196	–	1.5694	0.01427	–
12	6	8.8818	0.22636	1.14	0.4319	0.00823	0.42
12	8	4.4168	0.13008	0.79	0.1842	0.00416	0.80
r	R	ε_C	$\|C-C_r\|_{L^2(H^1)}$	Rate			
12	4	1.5529	0.02991	–			
12	6	0.6768	0.01468	0.85			
12	8	0.3858	0.00927	0.81			

2.5.3 *Test 2: Comparison of POD Solution and VMS-POD Solution*

In this test, we investigate the impact of the VMS method. We choose $\alpha_1 = \alpha_2 = \alpha_3 = 10^{-3}$, and $R = 5$. Figures 2.1, 2.2, 2.3 illustrate the decreasing behaviours of L^2 errors for each variable. It is clear from these figures that the VMS method

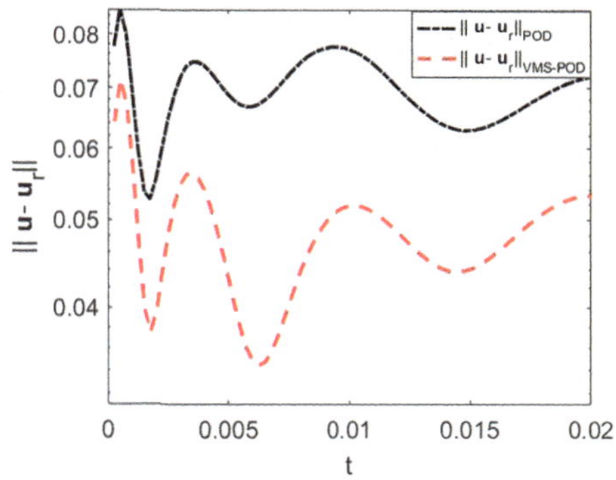

Fig. 2.1 L^2 errors for velocity of stabilized and unstabilized solution

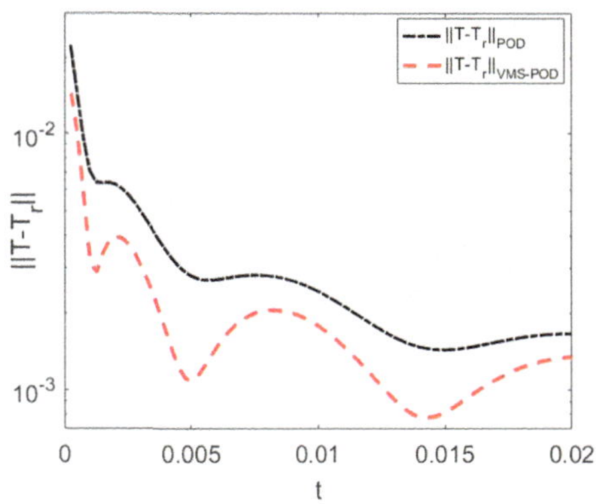

Fig. 2.2 L^2 errors for temperature of stabilized and unstabilized solution

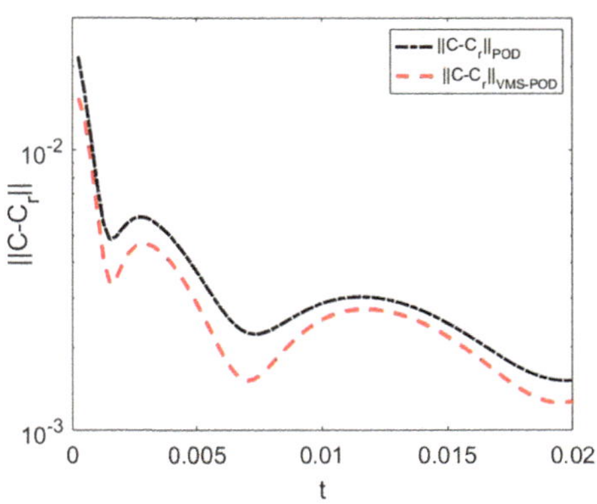

Fig. 2.3 L^2 errors for concentration of stabilized and unstabilized solution

improves the behaviour of the POD method. Hence, the VMS-POD model gives more accurate results than the POD model.

2.6 Conclusion

We propose an extrapolated VMS-POD method for Darcy Brinkman scheme. The algorithm (2.25)–(2.27) includes projection based VMS stabilization in POD for each fluid variable and treats the nonlinearity with the Crank Nicholson extrapolations. The numerical analysis of the proposed algorithm is performed. In addition, theoretical results are confirmed numerically and the efficiency of the algorithm (2.25)–(2.27) is presented.

References

1. Adams, R.A.: Sobolev Spaces. Academic Press, New York (1975)
2. Baker, G.: Galerkin Approximations for the Navier-Stokes Equations. Harvard University (1976)
3. Bennacer, R., Beji, H., Duval, R., Vasseur, P.: The Brinkman model for thermosolutal convection in a vertical annular porous layer. Int. Commun. Heat Mass **27**(1), 69–80 (2000)
4. Chen, S., Tölke, J., Krafczyk M.: Numerical investigation of double-diffusive (natural) convection in vertical annuluses with opposing temperature and concentration gradients. Int. J. Heat Fluid Flow **31**(2), 217–226 (2010)
5. Çıbık, A., Kaya, S.: Finite element analysis of a projection-based stabilization method for the Darcy-Brinkman equations in double-diffusive convection. Appl. Numer. Math. **64**, 35–49 (2013)
6. Eroglu, F., Kaya, S., Rebholz, L.: A modular regularized variational multiscale proper orthogonal decomposition for incompressible flows. Comput. Methods Appl. Mech. Eng. **325**, 350–368 (2017)
7. Eroglu, F., Kaya, S., Rebholz, L.: POD-ROM for the Darcy-Brinkman equations with double-diffusive convection. J. Numer. Math. (2018, to be published). Doi: https://doi.org/10.1515/jnma-2017-0122
8. Eroglu, F.: Reduced Order Modelling for Multiphysics problems. PhD thesis, Middle East Technical University, Turkey (2018)
9. Girault, V., Raviart, P.A.: Finite Element Approximation of the Navier-Stokes Equations. Lecture Notes in Math., vol. 749. Springer, Berlin (1979)
10. Goyeau, B., Songbe, J. P., Gobin, D.: Numerical study of double-diffusive natural convection in a porous cavity using the Darcy-Brinkman formulation. Int. J. Heat. Mass Tran. **39**(7), 1363–1378 (1996)
11. Iliescu, T., Wang, Z.: Variational multiscale proper orthogonal decomposition: convection-dominated convection-diffusion-reaction equations. Math. Comput. **82**(283), 1357–1378 (2013)
12. Iliescu, T., Wang, Z.: Variational multiscale proper orthogonal decomposition: Navier-Stokes equations. Numer. Methods Partial Differ. Equ. **30**(2), 641–663 (2014)
13. John, V., Kaya, S.: A finite element variational multiscale method for the Navier-Stokes equations. SIAM J. Sci. Comput. **26**, 1485–1503 (2005)

14. Karimi-Fard, M., Charrier-Mojtabi, M.C., Vafai, K.: Non-Darcian effects on double-diffusive convection within a porous medium. Numer. Heat Transf. Part A Appl. **31**(8), 837–852 (1997)
15. Kelliher, J.P., Temam, R., Wang, X.: Boundary layer associated with the Darcy-Brinkman-Boussinesq model for convection in porous media. Physica D **240**, 619–628 (2011)
16. Layton, W.: A connection between subgrid scale eddy viscosity and mixed methods. Appl. Math. Comput. **133**, 147–157 (2002)
17. Layton, W.: Introduction to the Numerical Analysis of Incompressible Viscous Flows. Society for Industrial and Applied Mathematics (SIAM), Philadelphia, USA (2008)
18. March, R., Coutinho, A.L.G.A., Elias, R.N.: Stabilized finite element simulation of double diffusive natural convection. Mecánica Computacional **XXIX**, 7985–8000 (2010)
19. Mojtabi, A., Charrier-Mojtabi, M.C.: Double-Diffusive Convection in Porous Media. Handbook of Porous Media, pp. 559–603. Taylor and Francis (2005)
20. Nield, D.A., Bejan, A.: Convection in Porous Media. Springer (1992)
21. Ravindran, S.: Real-time computational algorithm for optimal control of an MHD flow system. SIAM J. Sci. Comput. **26**(4), 1369–1388 (2005)
22. San, O., Borggaard, J.: Basis selection and closure for POD models of convection dominated Boussinesq flows. In: Proceeding of the Twenty-First International Symposium on Mathematical Theory of Networks and Systems, Groningen, The Netherlands, pp. 132–139 (2014)
23. San, O., Borggaard, J.: Principal interval decomposition framework for POD reduced-order modeling of convective Boussinesq flows. Int. J. Numer. Methods Fluids **78**, 37–62 (2015)
24. Sirovich, L.: Turbulence and the dynamics of coherent structures, Parts I, II and III. Q. Appl. Math. **45**, 561–590 (1987)
25. Stern, M.E.: The "salt-fountain" and thermohaline convection. Tellus **12**, 172–175 (1960)
26. Tsubaki, K., Maruyama, S., Komiya, A., Mitsugashira, H.: Continuous measurement of an artificial upwelling of deep sea water induced by the perpetual salt fountain. Deep Sea Res. Part I Oceanogr. Res. Pap. **54**, 75–84 (2007)

Chapter 3
Comparison of Exact and Numerical Solutions for the Sharma–Tasso–Olver Equation

Doğan Kaya, Asıf Yokuş, and Uğur Demiroğlu

3.1 Introduction

In this study, we implemented to find the exact solutions of STO equation [1–6] by using an aBTM [7–10] with the help of computer programming. In this application we establish the exact solutions of the STO equation

$$u_t + 3u_x^2 + 3u^2u_x + 3uu_{xx} + u_{xxx} = 0 \tag{3.1}$$

where $u = u(x, t)$ is a real function for all $x, t \in$ R. In literature, the STO equation has been studied in many applications by many physicists and mathematicians. Yan considered Eq. (3.1) by using the transformation of Cole–Hopf method [9]. Lian and Lou [2] have been implementing to get exact solutions of this equation by using the symmetry reduction scheme. Several papers of Eq. (3.1) are also published in the means of treating analytically by the various types of hyperbolic method (such as tanh and sech) and some ansatz consisting of hyperbolic and exponential functions in [11–13].

Nonlinear model of the equations for both mathematicians and physicists are very important to construct explicit and numerical solutions schemes. Because of the importance, in recent years, remarkable progress has been made in the

---

D. Kaya
Department of Mathematics, Istanbul Commerce University, Istanbul, Turkey
e-mail: dogank@ticaret.edu.tr

A. Yokuş (✉)
Department of Actuary, Firat University, Elazig, Turkey
e-mail: asfyokus@firat.edu.tr

U. Demiroğlu
Department of Computer Center, Firat University, Elazig, Turkey
e-mail: ugurdemiroglu@firat.edu.tr

© Springer Nature Switzerland AG 2020

J. A. T. Machado et al. (eds.), *Numerical Solutions of Realistic Nonlinear Phenomena*, Nonlinear Systems and Complexity 31, https://doi.org/10.1007/978-3-030-37141-8_3

establishment of the solutions for nonlinear either ordinary or partial differential equations (in short ODE or PDE) [8, 9]. It is very important to say that directly obtaining the solution of the nonlinear equation is physically meaningful because these types of solutions are keeping the actual physical characters [14]. There are many studies to create exact and numerical solutions for nonlinear PDEs [15–17]. To get the exact solution, some transformations are needed, such as Miura, Darboux, Cole–Hopf, the inverse scattering, and the Bäcklund transformation. The other commonly used methods are given as: tanh method, sine-cosine method, Painleve method, homogeneous balance method (HB), similarity reduction method, and so on [8, 9, 13]. On the other hand, some of the numerical methods [18–22] for the nonlinear PDE have been investigated such as the finite element method, Galerkin method, collocation methods with quadratic B-splines, an explicit multistep method, finite difference methods, Fourier Leap-Frog method, and some group of semi-analytic methods such as HAM, ADM, HPM, and so on [10].

3.2 Analysis of the Exact Solution Method

Now we will briefly give a description of the aBTM [7–10], for a given nonlinear PDE

$$\Psi\left(u, u_t, u_x, u_{xx}, \ldots\right) = 0. \tag{3.2}$$

We get the homogenous balance (HB) of Eq. (3.1) in the form

$$u = \partial_x^{\alpha} \partial_t^{\beta} f\left[w\right] + v, \tag{3.3}$$

where $w = w(x, t)$, $u = u(x, t)$, and $v = v(x, t)$ are undetermined functions and α, β are positive integers determined by balancing the highest derivative term with the nonlinear terms in Eq. (3.1) (see [7–10] for details). However, we find that the constants α, β should not be restricted to positive integers. Substituting Eq. (3.3) into Eq. (3.1) yields a set of algebraic equations for f', f'', ..., then all coefficients of these set are equal to zero. After the algebraic equation system is solved, we find the transformed form of Eq. (3.1); then by the solution of this transformed form we have the solutions of Eq. (3.1).

3.3 An Application of the Exact Solution Method

Let us consider Eq. (3.1). According to the idea of improved HB [7–10], we seek for Bäcklund transformation of Eq. (3.1). And according to the balancing principle by using Eq. (3.3) we have $\alpha = 1$ and $\beta = 0$. Therefore, we may choose

$$u = \partial_x f\,[w] + v = f' w_x + v, \tag{3.4}$$

where $f = f(w)$ and $w = w(x, t)$ are undetermined functions, and $u = u(x, t)$ and $v = v(x, t)$ are two solutions of Eq. (3.1). Substituting Eq. (3.4) into Eq. (3.1), we obtain

$$\begin{aligned}
&\left(3f'^2 f'' + 3f''^2 + 3f' f''' + f^{(4)}\right) w_x^4 + \left(3f'^3 + 15 f' f'' + 6f'''\right) w_x^2 w_{xx} \\
&+ \left(3vf''' + 6vf' f''\right) w_x^3 + (f'' w_{xt} + 3v^2 f'' w_x^2 + 6f'' v_x w_x^2 + 9vf'' w_x w_{xx} \\
&+ 3f'' w_{xx}^2 + 4f'' w_x w_{xxx} + 6vf'^2 w_x w_{xx} + 3f'^2 w_{xx}^2 + 3f'^2 w_x w_{xxx} + 3f'^2 v_x w_x^2) \\
&+ \left(f' w_{xt} + 3f' v_{xx} w_x + 3v^2 f' w_{xxx} + 6vf' v_x w_x\right) \\
&+ \left(v_t + 3v_x^2 + 3v^2 v_x + 3vv_{xx} + v_{xxx}\right) = 0.
\end{aligned} \tag{3.5}$$

Setting the coefficients of w_x^4 in Eq. (3.5) to zero, we obtain following differential equation:

$$3f'^2 f'' + 3f''^2 + 3f' f''' + f^{(4)} = 0, \tag{3.6}$$

which have solutions as in the following cases.

3.3.1 *Case 1*

If

$$f = 2 \ln w, \tag{3.7}$$

then by from Eq. (3.7) it holds that

$$f' f'' = -f''', \; f'^2 = -2f'', \; f'^3 = 2f'''. \tag{3.8}$$

By using nonlinear Eq. (3.8), Eq. (3.5) can be rewritten as the sum of some terms with $f', \; f'', \ldots$, equating the coefficients to zero yields

$$\begin{aligned}
(-3vw_x^3 - 3w_x^2 w_{xx}) f''' &= 0, \\
(w_t w_x + 3v^2 w_x^2 - 3vw_x w_{xx} - 3w_{xx}^2 - 2w_x w_{xxx}) f'' &= 0, \\
(6vv_x w_x + w_{xt} + 3w_x v_{xx} + 3v^2 w_{xx} + 6v_x w_{xx} + 3vw_{xxx} + w_{xxxx}) f' &= 0, \\
(v_t + 3v_x^2 + 3v^2 v_x + 3vv_{xx} + v_{xxx}) &= 0.
\end{aligned} \tag{3.9}$$

From the system Eq. (3.9) v can be taken as an arbitrary constant so that some of the terms are vanished. The coefficient of the derivative of f for the new system of

Eq. (3.10) will be

$$\begin{aligned}-3w_x^2(vw_x+w_{xx})&=0,\\ w_tw_x+3v^2w_x^2-3vw_xw_{xx}-3w_{xx}^2-2w_xw_{xxx}&=0,\\ \partial_x(w_t+3v^2w_x+3vw_{xx}+w_{xxx})&=0.\end{aligned} \tag{3.10}$$

In Eq. (3.9), the first equation $vw_x + w_{xx}$ is a linear PDE and so this equation has a type of solution as

$$w = a_0 + a_1 \exp[b(x - ct)], \tag{3.11}$$

where a_0, a_1, b, and c are arbitrary constants (Fig. 3.1). Substituting the excepted solution Eq. (3.11) into the set of Eq. (3.9), a set of nonlinear algebraic equations is obtained.

$$\begin{aligned}b+v&=0,\\ b^2-c+3bv+3v^2&=0,\\ 5b^2+c+3bv-3v^2&=0.\end{aligned} \tag{3.12}$$

From the equation system Eq. (3.12), we obtain

$$b = -\sqrt{c},\ v = \sqrt{c} \text{ and } b = \sqrt{c},\ v = -\sqrt{c}. \tag{3.13}$$

By means of computer program, substituting b in Eq. (3.13) into Eq. (3.11) and solution Eq. (3.11) into Eq. (3.7) we can find the wave solutions of Eq. (3.1). This solution is as following:

$$u_1(x,t) = -\sqrt{c} + \frac{a_1\sqrt{c}\left(\cosh\left[c^{3/2}t-\sqrt{c}x\right]-\sinh\left[c^{3/2}t-\sqrt{c}x\right]\right)}{a_0+a_1\cosh\left[c^{3/2}t-\sqrt{c}x\right]-a_1\sinh\left[c^{3/2}t-\sqrt{c}x\right]}, \tag{3.14}$$

$$u_2(x,t) = \sqrt{c} - \frac{a_1\sqrt{c}\left(\cosh\left[c^{3/2}t-\sqrt{c}x\right]+\sinh\left[c^{3/2}t-\sqrt{c}x\right]\right)}{a_0+a_1\cosh\left[c^{3/2}t-\sqrt{c}x\right]+a_1\sinh\left[c^{3/2}t-\sqrt{c}x\right]}. \tag{3.15}$$

3.3.2 *Case 2*

If

$$f = \ln w, \tag{3.16}$$

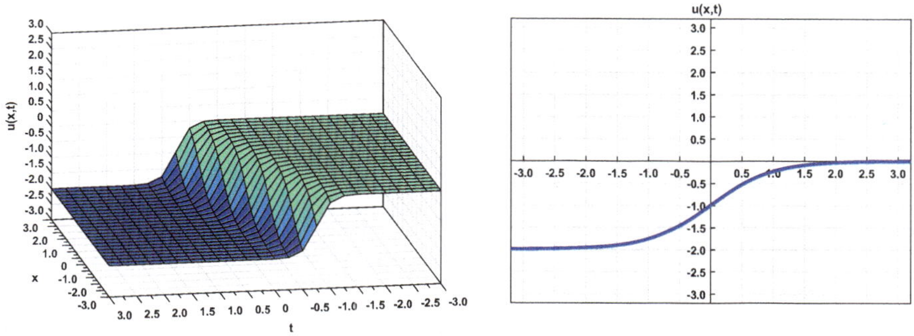

Fig. 3.1 Exact solution $u_1(x,t)$ of Eq. (3.14) by substituting the values $c = 4$, $a_0 = 1$, $a_1 = 1$, $-3 \le x \le 3$, $-3 \le t \le 3$, and $t = 0.00001$ for the 2D graphic for an aBTM application

then by from Eq. (3.16) it holds that

$$f' f'' = -\frac{1}{2} f''', \; f'^2 = -f'', \; f'^3 = \frac{1}{2} f'''. \tag{3.17}$$

Equation (3.5) can be rewritten as the sum of some terms with $f', \; f'', \ldots$ equating the coefficients to zero yields

$$\begin{aligned}(w_t w_x + 3v^2 w_x^2 + 3v_x w_x^2 + 3v w_x w_{xx} + w_x w_{xxx}) f'' &= 0,\\ (6v v_x w_x + w_{xt} + 3v_{xx} w_x + 3v^2 w_{xx} + 6v_x w_{xx} + 3v w_{xxx} + w_{xxxx}) f' &= 0,\\ \left(v_t + 3v_x^2 + 3v^2 v_x + 3v v_{xx} + v_{xxx}\right) &= 0.\end{aligned} \tag{3.18}$$

From the system Eq. (3.18) v can be taken as an arbitrary constant and where the third equation of the system Eq. (3.18) gives $w_t + 3v^2 w_x + 3v w_{xx} + w_{xxx} = 0$, which is a linear PDE and so this equation has a solution as

$$w = a_0 + a_1 \exp[b(x - ct)], \tag{3.19}$$

where a_0, a_1, b, and c are arbitrary constants (Fig. 3.2). Substituting Eq. (3.19) into the set of Eq. (3.10), a set of nonlinear algebraic equations yields

$$b^2 - c + 3bv + 3v^2 = 0. \tag{3.20}$$

From the algebraic Eq. (3.20) we obtain

$$b = \frac{1}{2}\left(-3v + \sqrt{4c - 3v^2}\right) \text{ and } b = \frac{1}{2}\left(-3v - \sqrt{4c - 3v^2}\right). \tag{3.21}$$

By means of Mathematica, substituting Eq. (3.21) into Eq. (3.19) and Eq. (3.19) into Eq. (3.4) we have the following kink-type solutions of Eq. (3.10):

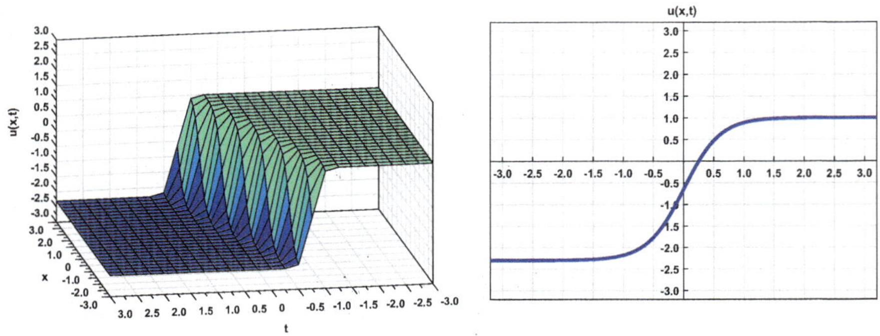

Fig. 3.2 Exact solution $u_1(x,t)$ of Eq. (3.22) by substituting the values $c = 4$, $a_0 = 1$, $a_1 = 1$, $v = 1$, $-3 \leq x \leq 3$, $-3 \leq t \leq 3$, and $t = 0.00001$ for the 2D graphic for an aBTM application

$$u_1(x,t) = v + \frac{a_1\left(-3v - \sqrt{4c-3v^2}\right)\begin{bmatrix}\cosh\left[\frac{1}{2}\left(-3v-\sqrt{4c-3v^2}\right)(-ct+x)\right]\\ +\sinh\left[\frac{1}{2}\left(-3v-\sqrt{4c-3v^2}\right)(-ct+x)\right]\end{bmatrix}}{2\begin{bmatrix}a_0 + a_1\cosh\left[\frac{1}{2}\left(-3v-\sqrt{4c-3v^2}\right)(-ct+x)\right]\\ +a_1\sinh\left[\frac{1}{2}\left(-3v-\sqrt{4c-3v^2}\right)(-ct+x)\right]\end{bmatrix}}, \tag{3.22}$$

$$u_2(x,t) = v + \frac{a_1\left(-3v + \sqrt{4c-3v^2}\right)\begin{bmatrix}\cosh\left[\frac{1}{2}\left(-3v+\sqrt{4c-3v^2}\right)(-ct+x)\right]\\ +\sinh\left[\frac{1}{2}\left(-3v+\sqrt{4c-3v^2}\right)(-ct+x)\right]\end{bmatrix}}{2\begin{bmatrix}a_0 + a_1\cosh\left[\frac{1}{2}\left(-3v+\sqrt{4c-3v^2}\right)(-ct+x)\right]\\ +a_1\sinh\left[\frac{1}{2}\left(-3v+\sqrt{4c-3v^2}\right)(-ct+x)\right]\end{bmatrix}}. \tag{3.23}$$

3.4 Analysis of Finite Difference Method

The following notations are needed to express some for using numerical FDM:

1. A corresponding choice of the time and spatial steps Δt and Δx, respectively,
2. The coordinates of mesh points are given as $x_i = a + i\Delta x$ and $t_j = j\Delta t$, $i; j = 0, 1, 2, \ldots, N; M$ where $N = \frac{b-a}{\Delta x}$ and $M = \frac{T}{\Delta t}$, respectively.
3. The solution of the given function $u(x,t)$ can be written their grid points as $u(x_i, t_j) \cong u_{i,j}$ which are represent the numerical results of the values of $u(x,t)$ at the points of (x_i, t_j).

We have the difference operators as [18]

$$H_t u_{i,j} = u_{i,j+1} - u_{i,j}, \tag{3.24}$$

$$H_x u_{i,j} = u_{i+1,j} - u_{i,j}, \tag{3.25}$$

$$H_{xx} u_{i,j} = u_{i+1,j} - 2u_{i,j} + u_{i-1,j}, \tag{3.26}$$

$$H_{xxx} u_{i,j} = u_{i+2,j} - 2u_{i+1,j} + 2u_{i-1,j} - u_{i-2,j}. \tag{3.27}$$

Therefore corresponding terms of the derivatives of the given equation will be replaced with the discrete operators for FDM as following:

$$\left.\frac{\partial u}{\partial t}\right|_{i,j} = \frac{H_t u_{i,j}}{\Delta t} + O(\Delta t), \tag{3.28}$$

$$\left.\frac{\partial u}{\partial x}\right|_{i,j} = \frac{H_x u_{i,j}}{\Delta x} + O(\Delta x), \tag{3.29}$$

$$\left.\frac{\partial^2 u}{\partial x^2}\right|_{i,j} = \frac{H_{xx} u_{i,j}}{(\Delta x)^2} + O(\Delta x^2), \tag{3.30}$$

$$\left.\frac{\partial^3 u}{\partial x^3}\right|_{i,j} = \frac{H_{xxx} u_{i,j}}{2\,(\Delta x)^3} + O(\Delta x^2), \tag{3.31}$$

initial values $u_{i,0} = u_0(x_i)$.

3.4.1 Truncation Error and Stability Analysis

In this part of the section, an investigation of the stability and error analysis of the FDE will be given. With the classical definition of the stability, if there is a small change in the initial condition, then this change would not cause a large error in the final numerical results.

Theorem 3.1 *If the truncation of the finite difference formula of error to the STO equation is E, then* $\lim\limits_{\substack{\Delta x \to 0 \\ \Delta t \to 0}} E = 0$.

Proof Substituting Eqs. (3.28)–(3.31) into Eq. (3.1) gives

$$\begin{aligned} &\left(\frac{H_t u_{i,j}}{\Delta t} + O(\Delta t)\right) + 3\left(\frac{H_x u_{i,j}}{\Delta x} + O(\Delta x)\right)^2 + 3u_{i,j}^2\left(\frac{H_x u_{i,j}}{\Delta x} + O(\Delta x)\right) \\ &+ 3u_{i,j}\left(\frac{H_{xx} u_{i,j}}{(\Delta x)^2} + O(\Delta x^2)\right) + \frac{H_{xxx} u_{i,j}}{2\,(\Delta x)^3} + O(\Delta x^2) = 0. \end{aligned} \tag{3.32}$$

We could rewrite Eq. (3.32) in a suitable way; final formulation could be arranged in the following form:

$$\left(\frac{H_t u_{i,j}}{\Delta t}\right)+3\left(\frac{H_x u_{i,j}}{\Delta x}\right)^2+3u_{i,j}^2\left(\frac{H_x u_{i,j}}{\Delta x}\right)+3u_{i,j}\left(\frac{H_{xx} u_{i,j}}{(\Delta x)^2}\right)+\frac{H_{xxx} u_{i,j}}{2\,(\Delta x)^3}$$
$$+\,O(\Delta x^2)+O(\Delta t)+3\,(O(\Delta x))^2+6\frac{H_x u_{i,j}}{\Delta x}O(\Delta x)+3u_{i,j}^2 O(\Delta x)$$
$$+\,3u_{i,j}O(\Delta x^2)=0. \tag{3.33}$$

If we expand the truncation error, we get below equation

$$E=O(\Delta x^2)+O(\Delta t)+3\,(O(\Delta x))^2+6\frac{H_x u_{i,j}}{\Delta x}O(\Delta x)+3u_{i,j}^2 O(\Delta x)+3u_{i,j}O(\Delta x^2), \tag{3.34}$$

after the expanding and separating Eq. (3.32) will get the indexed of Eq. (3.1),

$$\left(\frac{H_t u_{i,j}}{\Delta t}\right)+3\left(\frac{H_x u_{i,j}}{\Delta x}\right)^2+3u_{i,j}^2\left(\frac{H_x u_{i,j}}{\Delta x}\right)+3u_{i,j}\left(\frac{H_{xx} u_{i,j}}{(\Delta x)^2}\right)+\frac{H_{xxx} u_{i,j}}{2\,(\Delta x)^3}=0. \tag{3.35}$$

If we substitute the equalities Eqs. (3.24)–(3.27) into Eq. (3.35) and then do some algebraic manipulations, the following equality will be constructed:

$$u_{i+1,j}=\frac{1}{6\Delta x\sqrt{\Delta t}}\left\{\begin{array}{l}\sqrt{\Delta t}\,(1-3\Delta x u_{i,j}\,(-2+\Delta x u_{i,j}))\\ -\left[\begin{array}{l}\Delta t+6\Delta t\Delta x u_{i-2,j}-12\Delta t\Delta x u_{i-1,j}\,(1+3\Delta x u_{i,j})\\ +3\Delta x\left(\begin{array}{l}22\Delta t\Delta x u_{i,j}^2+3\Delta t\,(\Delta x)^3 u_{i,j}^4\\ +4u_{i,j}\left(\Delta t+(\Delta x)^3-3\Delta t\Delta x u_{i,j+1}\right)\\ -2\left(2\,(\Delta x)^3 u_{i,j+1}+\Delta t u_{i+2,j}\right)\end{array}\right)\end{array}\right]^{\frac{1}{2}}\end{array}\right\}. \tag{3.36}$$

By using Eq. (3.36), we can write numerical solution $\hat{U}$ as $\hat{U}=u_{i+1,j}$. Moreover, E transaction error could be written $E=\left|U-\hat{U}\right|$, where U is found exact and $\hat{U}$ is a corresponding numerical solution. Obviously, using Eq. (3.33) we can conclude that if Δx and Δt are taking as small as necessary E error will be very small. As result of this, limit of E would be written

$$\lim_{\substack{\Delta x\to 0\\ \Delta t\to 0}} E=0. \tag{3.37}$$

Now, if Δt and Δx are taken as small as close to zero $\varepsilon>0$, then above equality will be written $|E|<\varepsilon$. From this expression, we could say that FDM is stable.

Theorem 3.2 *The FDM for the STO equation is linear stable.*

Proof We consider the von Neumann's stability of the FDM for the STO equation [22]. Let

$$u_{i,j}=u(i\Delta x,j\Delta t)=u(p,q)=\varepsilon^q e^{I\xi p},\ \xi\in[-\pi,\pi], \tag{3.38}$$

where $p = i\Delta x$, $q = j\Delta t$, and $I = \sqrt{-1}$. Here, we examine the stability of the numerical scheme with STO equation with strong nonlinearity by using the Fourier–von Neumann stability analysis. To carry out this analysis, we linearize the nonlinear terms $3u_x^2$, $3u^2u_x$, and $3uu_x$ by taking $\widehat{u_1} = 3u_x$, $\widehat{u_2} = 3u^2$ and $\widehat{u_3} = 3u$ as local constants. Therefore, the nonlinear terms $3u_x^2$, $3u^2u_x$, and $3uu_{xx}$ in the equation were changed to $\widehat{u_1}u_x$, $\widehat{u_2}u_x$, and $\widehat{u_3}u_{xx}$, respectively,

$$u_t + \widehat{u_1}u_x + \widehat{u_2}u_x + \widehat{u_3}u_{xx} + u_{xxx} = 0. \tag{3.39}$$

Inserting Eqs. (3.28)–(3.31) into Eq. (3.39) yields

$$\xi = X - iY, \tag{3.40}$$

where

$$X = \frac{1}{((\Delta x)^2 + (\Delta t)\widehat{u_3})}\begin{pmatrix} (\Delta x)^2 + (\Delta x)(\Delta t)\widehat{u_1} + (\Delta x)(\Delta t)\widehat{u_2} + 2(\Delta t)\widehat{u_3} \\ - (\Delta x)(\Delta t)\widehat{u_1}\cos(\phi) - (\Delta x)(\Delta t)\widehat{u_2}\cos(\phi) \\ - (\Delta t)\widehat{u_3}\cos(\phi) \end{pmatrix} \tag{3.41}$$

and

$$Y = \frac{(\Delta t)\sin(\phi)}{(\Delta x)((\Delta x)^2 + (\Delta t)\widehat{u_3})}\left(-2 + (\Delta x)^2\widehat{u_1} + (\Delta x)^2\widehat{u_2} - (\Delta x)\widehat{u_3} + 2\cos(\phi)\right). \tag{3.42}$$

By the Fourier stability, for a given numerical scheme to be stable, $|\xi| \leq 1$ and $|\xi|^2 = X^2 + Y^2$ must be satisfied. The stability of the numerical scheme depends on choosing the values of (Δx) and (Δt) to be very small.

3.4.2 L_2 and L_∞ Error Norms

For the considered model Eq. (3.1) of the approximate numerical solutions is also calculated with the help of the Mathematica programming environments. In order to conclude that how the explicit and corresponding approximate results are how much close to each, we use error norms similarly defined in reference [18] as

$$L_2 = \left\| u^{exact} - u^{numeric} \right\|_2 = \sqrt{h \sum_{j=0}^{N} \left| u_j^{exact} - u_j^{numeric} \right|^2}$$

and

$$L_\infty = \left\| u^{exact} - u^{numeric} \right\|_\infty = \underset{j}{Max} \left| u_j^{exact} - u_j^{numeric} \right|.$$

Table 3.1 Exact and corresponding numerical solutions of Eq. (3.1) and its absolute errors with $\Delta x = 0.00001$ and $0 \leq x \leq 1$

x_i	t_j	$NumericalSolution$	$ExactSolution$	$Error$
0.00000	0.00001	−0.9999899998369807	−1.0000399999999785	0.000050000162997831940
0.00001	0.00001	−0.9999799998295793	−1.0000299999999909	0.000050000170411568234
0.00002	0.00001	−0.9999699998295847	−1.0000199999999975	0.000050000170412789480
0.00003	0.00001	−0.9999599998406973	−1.0000099999999996	0.000050000159302343580
0.00004	0.00001	−0.9999499998296136	−1.0000000000000000	0.000050000170386366170
0.00005	0.00001	−0.9999399998333433	−0.9999900000000002	0.000050000166656904990
0.00006	0.00001	−0.9999299998333838	−0.9999800000000028	0.000050000166618935360

3.5 Implementation of the FDEs

Here, the presentation of the numerical results have been obtained by using the following data: $c = 4$, $a_0 = 1$, $a_1 = 1$, $0 < x < 1$, and $0 < t < 1$ for Eq. (3.14), the initial conditions are

$$u_0(x) = u_1(x,0) = -2 + \frac{2\left(\cosh[2x] + \sinh[2x]\right)}{1 + \cosh[2x] + \sinh[2x]} = -1 + \tanh[x] \tag{3.43}$$

and using the above suppositions, the explicit solution of Eq. (3.1) will be written as

$$\begin{aligned} u_1(x,t) &= -2 + \frac{2\cosh[8t-2x]}{1+\cosh[8t-2x]-\sinh[8t-2x]} \\ &\quad - \frac{2\sinh[8t-2x]}{1+\cosh[8t-2x]-\sinh[8t-2x]} \\ &= -\frac{2}{1+\cosh[8t-2x]-\sinh[8t-2x]}. \end{aligned} \tag{3.44}$$

According to the finite forward difference method, inserting the values $(\Delta x) = (\Delta t) = 0.0000001$ into Eq. (3.14), yields the following:

$$u_{i+1,j} = 4.166 \times 10^{-12} \left(\begin{array}{l} 3.999 \times 10^{15} + 2.4 \times 10^{11} u_{i,j} - 1.2 \times 10^{6} u_{i,j}^{2} \\ \quad - 1.732 \left[\begin{array}{l} 5.333 \times 10^{30} + 3.1999 \times 10^{26} u_{i-2,j} \\ - 6.399 \times 10^{26} u_{i-1,j} + 6.4 \times 10^{26} u_{i,j} \\ - 1.92 \times 10^{22} u_{i-1,j} u_{i,j} + 3.52 \times 10^{22} u_{i,j}^{2} \\ + 4.8 \times 10^{11} u_{i,j}^{4} - 6.4 \times 10^{16} u_{i,j+1} \\ - 1.92 \times 10^{22} u_{i,j} u_{i,j+1} - 3.199 \times 10^{26} u_{i+2,j} \end{array} \right]^{\frac{1}{2}} \end{array} \right).$$

We compare exact and numerical solutions in Tables 3.1 and 3.2.

Table 3.2 L_2 and L_∞ error norm when $0 \leq \Delta x \leq 1$ and $0 \leq \Delta t \leq 1$

$\Delta x = \Delta t$	L_2	L_∞
0.05000	0.2033790000	0.2531790000
0.01000	0.0395503000	0.0501589000
0.00500	0.0196868000	0.0250407000
0.00100	0.0039227400	0.0050016600
0.00010	0.0001574410	0.0005000170
0.00001	0.0000157575	0.0000500002

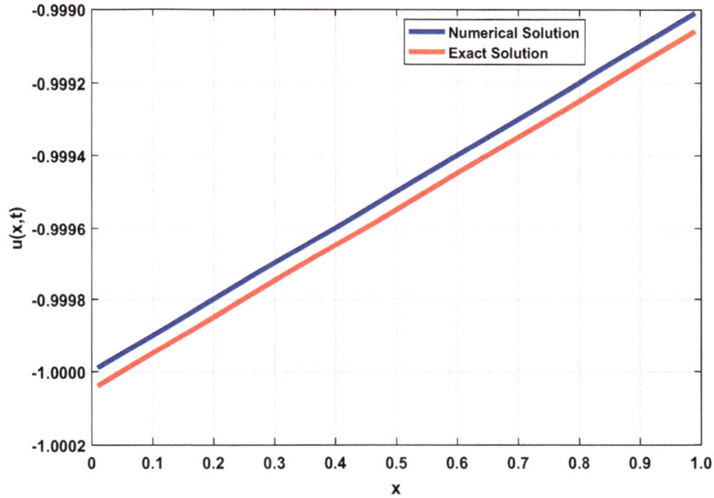

Fig. 3.3 A comparison for exact and corresponding numerical solutions of Eq. (3.1)

The L_2 and L_∞ error norm is shown in Table 3.2.
We could conclude that explicit and corresponding approximate solutions are in good agreement which are illustrated above numerical results are Tables 3.1 and 3.2.

Figures 3.3, 3.4 shows that the exact and corresponding numerical solutions of Eq. (3.1) are very close results which desired result. Because, our considered numerical method is stable and truncation error due to very much the choice of the Δx and Δt. This conclusion of the behavior of the exact and corresponding approximate solutions can be seen in the following depicted graph for the special value of $\Delta x = 0.00001$.

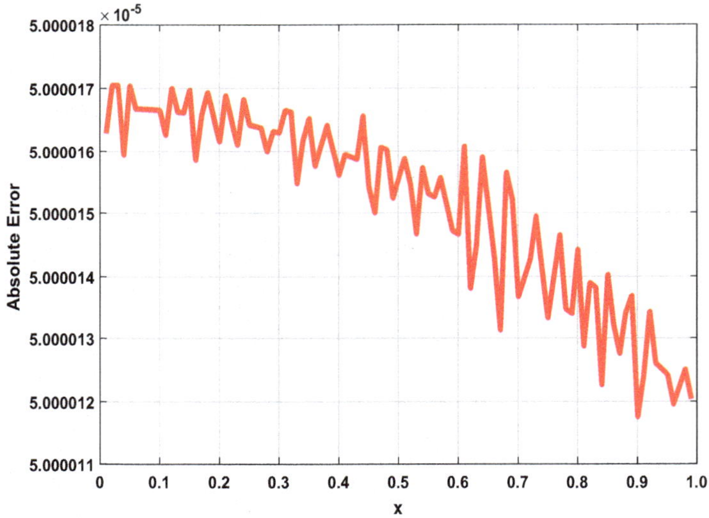

Fig. 3.4 The behavior of the exact and corresponding approximate solutions for absolute error of Eq. (3.1)

3.6 Conclusions

In this paper, we present aBT method which is applied to integrable STO equation. Using this transformation method it is easy to see that the nonlinear PDE such as STO equations transforming to linear PDEs. Thus we can obtain two traveling wave solutions of this equation. We present the plotting 2D and 3D surfaces to these obtained solutions. This method of application is also easy to build programming based on algebraic programming. Here important to point out that if the taken equation is not integrable (e.g., Painleve integrable, C-integrable, S-integrable, Lax integrable, and Liouville integrable, etc.) this algorithm of method is not working.

Acknowledgements The authors of the article would like to thank Istanbul Commerce University and Firat University for supporting this work.

References

1. Yan, Z.: Integrability of two types of the (2+1)-dimensional generalized Sharma-Tasso-Olver integro-differential equations. MM Res. **22**, 302–324 (2003)
2. Lian, Z., Lou, S.Y.: Symmetries and exact solutions of the Sharma-Tasso-Olver equation. Nonlinear Anal. **63**, 1167–1177 (2005)
3. Wang, S., Tang, X., Lou, S.Y.: Soliton fission and fusion: Burgers equation and Sharma-Tasso-Olver equation. Chaos Solitons Fractals **21**, 231–239 (2004)

4. Wazwaz, A.M.: New solitons and kinks solutions to the Sharma-Tasso-Olver equation. Appl. Math. Comput. **188**, 1205–1213 (2007)
5. Inan, I.E., Kaya, D.: Exact solutions of some nonlinear partial differential equations. Physica A **381**, 104–115 (2007)
6. Shang, Y., Qin, J., Huang, Y., Yuan, W.: Abundant exact and explicit solitary wave and periodic wave solutions to the Sharma-Tasso-Olver equation. Appl. Math. Comput. **202**, 532–538 (2008)
7. Fan, E.G.: Two new applications of the homogeneous balance method. Phys. Lett. A **265**, 353–357 (2000)
8. Ablowitz, M.J., Clarkson, P.A.: Solitons, Nonlinear Evolution Equations and Inverse Scattering. Cambridge University Press, Cambridge, UK (1991)
9. Drazin, P.G., Johnson, R.S.: Solitons: An Introduction. Cambridge Texts in Applied Mathematics. Cambridge University Press, Cambridge, UK (1989)
10. Fan, E.G.: Auto-Bäcklund transformation and similarity reductions for general variable coefficient KdV equations. Phys. Lett. A **265**, 26–30 (2002)
11. Fan, E.G.: Uniformly constructing a series of explicit exact solutions to nonlinear equations in mathematical physics. Chaos Solitons Fractals **16**, 819–839 (2003)
12. Shang, Y.D.: The extended hyperbolic function method and exact solutions of the long-short wave resonance equations. Chaos Solitons Fractals **36**, 762–771 (2008)
13. Wazwaz, A.M.: Partial Differential Equations: Methods and Applications. Balkema, Rotterdam (2002)
14. Hu, X.B., Ma, W.X.: Application of Hirota's bilinear formalism to the Toeplitz lattice-some special soliton-like solutions. Phys. Lett. A **293**, 161–165 (2002)
15. Abourabia, A.M., El Horbaty, M.M.: On solitary wave solutions for the two-dimensional nonlinear modified Kortweg-de Vries-Burger equation. Chaos Solitons Fractals **29**, 354–364 (2006)
16. Yavuz, M., Ozdemir, N.: Comparing the new fractional derivative operators involving exponential and Mittag-Leffler kernel. Discrete Continuous Dyn. Syst., 1098–1107 (2019)
17. Gulbahar, S., Yokus, A., Kaya, D.: Numerical solutions of Fisher's equation with collocation method. In: AIP Conference Proceedings, Vol. 1676(1), p. 020099. AIP Publishing (2015)
18. Yokus, A., Kaya, D.: Numerical and exact solutions for time fractional Burgers equation. J. Nonlinear Sci. Appl., **10**, 3419–3428 (2017)
19. Yuste, S.B.: Weighted average finite difference methods for fractional diffusion equations. J. Comput. Phys. **216**, 264–274 (2006)
20. Yokus, A.: Numerical Solutions of Time Fractional Korteweg–de Vries Equation and Its Stability Analysis. Communications Faculty of Sciences University of Ankara Series A1 Mathematics and Statistics, vol. 68(1), pp. 353–361
21. Yavuz, M., Ozdemir, N.: Numerical inverse Laplace homotopy technique for fractional heat equations. Therm. Sci. **22**(1), 185–194 (2018)
22. Guo, B., Pu, X., Huang, F.: Fractional Partial Differential Equations and Their Numerical Solutions. South China University of Technology, China (2015)

Chapter 4
A Linear B-Spline Approximation for a Class of Nonlinear Time and Space Fractional Partial Differential Equations

Behrouz Parsa Moghaddam, José António Tenreiro Machado, and Arman Dabiri

4.1 Introduction

Recent studies have shown that fractional partial differential equations (FPDEs) are superior to model anomalous diffusion processes [2, 3, 12, 16, 26, 33]. Following these ideas, the fractional Kolmogorov–Petrovskii–Piskunov, Newell–Whitehead–Segel, FitzHugh–Nagumo, and Fisher equations were proposed for the diffusion equation with a nonlinear source term. FPDEs emerge in a wide variety of applications like electromagnetic, acoustics, electrochemistry, cosmology, and material science [18, 23, 25, 27, 29]. Numerical solution algorithms for solving FPDEs are still at their early stage of development. Nonetheless, FPDEs received considerable attention during the past years and we can find the approximated solution of space-fractional diffusion equations using finite difference methods [15, 17, 21, 32, 33], finite element methods [7, 8, 24, 28, 30, 31], and spectral [1] methods.

In this chapter, FPDEs are solved using different numerical methods. As for any numerical algorithm, it is crucial to study the effects of the round-off error on the sensitivity of the algorithms. Spline functions overcome this issue, as they are commonly used to construct stable numerical methods. For this purpose, we employ a linear B-spline interpolation for the spatial discretization and an upwind finite difference method for the time discretization.

B. P. Moghaddam (✉)
Department of mathematics, Lahijan Branch, Islamic Azad University, Lahijan, Iran
e-mail: parsa@liau.ac.ir

J. A. Tenreiro Machado
Institute of Engineering, Polytechnic of Porto, Porto, Portugal

A. Dabiri
School of Engineering Technology, Eastern Michigan University, Ypsilanti, MI, USA

© Springer Nature Switzerland AG 2020

J. A. T. Machado et al. (eds.), *Numerical Solutions of Realistic Nonlinear Phenomena*, Nonlinear Systems and Complexity 31, https://doi.org/10.1007/978-3-030-37141-8_4

The outline of this chapter is as follows. In Sect. 4.2, the space-fractional integral and time-fractional derivative operators are discretized. In Sect. 4.3, two unconditionally stable methods for solving a class of space- and time-FPDEs with one space variable are presented. The error analysis and the stability of the numerical methods are also discussed. In Sect. 4.4, the efficiency and accuracy of the present approaches are examined for several special cases of the fractional KPP equation with the initial-boundary value conditions. Finally, Sect. 4.5 draws the main conclusions.

4.2 Discretization of the Fractional Operators

In this section, a brief review on the fractional-order operators is given. Moreover, a technique for the discretization of the space-fractional integral and time-fractional derivative operators is also presented.

Several types of fractional-order integral and derivative operators have been proposed based on different concepts [14].

Definition 4.2.1 The fractional-order integration operator in the sense of Riemann–Liouville is defined by

$$\mathscr{I}_{0,t}^{\alpha} u(x,t) := \frac{1}{\Gamma(\alpha)} \int_0^t (t-\zeta)^{\alpha-1} u(x,\zeta) \mathrm{d}\zeta, \qquad \alpha > 0, \tag{4.1}$$

where ζ is an auxiliary variable belonging to the interval $(0, t)$ and $\Gamma(\cdot)$ is the gamma function. Moreover, for a smooth function $u(x, t)$, we have [4]

$$\mathscr{I}_{0,t}^{\alpha}\, u_t^{(\alpha)}(x,t) = u(x,t) + \sum_{j=0}^{n-1} \frac{t^j}{j!} u_t^{(j)}(x,0), \tag{4.2}$$

where $n - 1 < \alpha \leq n, n \in \mathbb{N}$.

Definition 4.2.2 Let us consider $n - 1 < \alpha < n$, $n \in \mathbb{N}$, a function $u(x, t)$ that is $n - 1$ times continuously differentiable with respect to t and $u_t^{(n)}(x, t)$ that is at least once integrable. Then, the fractional-order derivative operator in the sense of Caputo is defined by

$$u_t^{(\alpha)}(x,t) := \frac{1}{\Gamma(n-\alpha)} \int_0^t \frac{u_t^{(n)}(x,\zeta)}{(t-\zeta)^{\alpha+1-n}} \mathrm{d}\zeta. \tag{4.3}$$

Proposition 4.2.1 *Suppose that* $u(x,t) \in C^2(\Omega)$, $\Omega = (0, L) \times (0, T) \subseteq \mathbb{R}^2$, *and let us define the space and time grid points by*

$$x_i = i\, h_x, \quad i = 0, 1, \ldots, N, \tag{4.4a}$$

$$t_k = k\,h_t, \quad k = 0, 1, \ldots, M, \tag{4.4b}$$

where $h_x = \frac{L}{N}$ *and* $h_t = \frac{T}{M}$. *Then, an approximation of the* α*-order spatial integral of* $u(x,t)$ *at the space point* $x = x_m$ *is*

$$\mathscr{I}^{\alpha}_{0,x_m} u(x,t_n) = \sum_{k=0}^{m} \frac{h_x^{\alpha}}{\Gamma(2+\alpha)} a_{m,k} u(x_k,t_n) + \mathscr{E}_1(x_m,t_n), \tag{4.5}$$

so that

$$a_{m,k} = \begin{cases} m^{\alpha}(\alpha+1-m) + (m-1)^{\alpha+1}, & k = 0 \\ (m-k-1)^{\alpha+1} + (m-k+1)^{\alpha+1} - 2(m-k)^{\alpha+1}, & 1 \le k \le m-1, \\ 1, & k = m \end{cases} \tag{4.6}$$

and the ℓ_∞*-norm of the approximation error* $\mathscr{E}_1(x_m,t_n)$ *is bounded as*

$$\|\mathscr{E}_1(x_m,t_n)\|_\infty \le \frac{m^{\alpha} h_x^{2+\alpha}}{\Gamma(\alpha+1)} \left\| u_x^{(2)}(x_m,t_n) \right\|_\infty. \tag{4.7}$$

Proof We build the piecewise linear interpolant of $u(x,t)$ in the spatial space x upon the interpolate of the given function on each sub-interval $[x_k, , x_{k+1}]$ with the first-degree polynomial:

$$u(x,t) \approx s_m(x,t) = \sum_{k=0}^{m-1} \left(\frac{x - x_{k+1}}{x_k - x_{k+1}} u(x_k,t) + \frac{x - x_k}{x_{k+1} - x_k} u(x_{k+1},t) \right). \tag{4.8}$$

Then, the substitution of the piecewise linear interpolant (4.8) into (4.1) yields

$$\begin{aligned} \mathscr{I}^{\alpha}_{0,x_m} u(x,t_n) &\approx \mathscr{I}^{\alpha}_{0,x_m} s_m(x,t_n) \\ &= \frac{1}{\Gamma(\alpha)} \sum_{k=0}^{m-1} \left\{ \int_{x_k}^{x_{k+1}} (x_m - \zeta)^{\alpha-1} s_m(\zeta,t_n) \mathrm{d}\zeta \right\} \\ &= \frac{1}{\Gamma(\alpha)} \sum_{k=0}^{m-1} \left(\int_{x_k}^{x_{k+1}} (x_m - \zeta)^{\alpha-1} \frac{\zeta - x_{k+1}}{x_k - x_{k+1}} \mathrm{d}\zeta \right) u(x_k,t_n) \\ &+ \frac{1}{\Gamma(\alpha)} \sum_{k=0}^{m-1} \left(\int_{x_k}^{x_{k+1}} (x_m - \zeta)^{\alpha-1} \frac{\zeta - x_k}{x_{k+1} - x_k} \mathrm{d}\zeta \right) u(x_{k+1},t_n). \end{aligned} \tag{4.9}$$

Finally, the integral in the summation of (4.9) can be calculated explicitly knowing that $x_k = k\,h_x$. This obtains the coefficients (4.6) and completes the proof.

Proposition 4.2.2 *Let $u(x,t) \in C^2(\Omega)$ and $0 < \alpha \leq 1$. Then, the α-order temporal derivative of $u(x,t)$ at $x = x_m$ is approximated by*

$$u_t^{(\alpha)}(x_m, t) = \sum_{k=0}^{n} \frac{h_t^{1-\alpha}}{\Gamma(3-\alpha)} b_{k,n} \delta u(x_m, t_k) + \mathscr{E}_2(x_m, t_n), \tag{4.10}$$

and the corresponding error $\mathscr{E}_2(x_m, t_n)$ at t_n is bounded by

$$\|\mathscr{E}_2(x_m, t_n)\|_\infty \leq \frac{2h_t^{2-\alpha}}{\Gamma(2-\alpha)} \left\| u_t^{(2)}(x_m, t_n) \right\|_\infty, \tag{4.11}$$

where $\delta u(x, t_k)$ is the central differential operator and

$$b_{k,n} = \begin{cases} (n-1)^{2-\alpha} - n^{1-\alpha}(n-2+\alpha), & k = 0 \\ (n-k+1)^{2-\alpha} - 2(n-k)^{2-\alpha} + (n-k-1)^{2-\alpha}, & 1 \leq k \leq n-1. \\ 1, & k = n \end{cases} \tag{4.12}$$

Proof The proof is similar to that of Proposition 4.2.1 by using the central finite difference approximation for the time-fractional derivative.

4.3 Numerical Approach for the FPDE

In this section, we consider a class of the space- and time-FPDEs.

4.3.1 Space-FPDE

Consider the following space-FPDE with the solution $u(x,t) \in C^2(\Omega)$, $\Omega = (0, L) \times (0, T) \subseteq \mathbb{R}^2$,

$$u_x^{(\alpha)}(x,t) = u_t^{(1)}(x,t) + f(t, x, u(x,t)), \tag{4.13}$$

subject to the initial and boundary conditions

$$u(x, 0) = q(x), \tag{4.14a}$$

$$u_x^{(j)}(0, t) = p_j(t), \qquad j = 0, 1, \ldots, n-1, \tag{4.14b}$$

where $n - 1 < \alpha \leq n$, $n \in \mathbb{N}$.

We assume that $p_j(t)$ and $q(x)$ are smooth functions, and the nonlinear source term $f(t,x,u)$ satisfies the Lipschitz condition with the Lipschitz constant $l_f > 0$:

$$|f(t,x,u) - f(t,x,w)| \leq l_f |u - w|. \tag{4.15}$$

In addition, let u_m^n denote the numerical solution of (4.13) at the grid point (x_m, t_n) with the mesh step sizes h_x and h_t.

According to Definition 4.2.1, the integral form of the initial boundary value problem (4.13) is

$$u(x,t) = \sum_{j=0}^{n-1} \frac{x^j}{j!} u_x^{(j)}(0,t) + \mathscr{I}_{0,x}^{\alpha} \left(u_t^{(1)}(x,t) + f(t,x,u(x,t)) \right), \tag{4.16}$$

with solution at the point (x_m, t_n) given by

$$u(x_m,t_n) = \sum_{j=0}^{n-1} \frac{x_m^j}{j!} u_x^{(j)}(0,t_n) + \mathscr{I}_{0,x_m}^{\alpha} \left(u_t^{(1)}(x,t_n) + f(t_n,x,u(x,t_n)) \right). \tag{4.17}$$

The fractional integral is approximated using Proposition 4.2.1 and the formula (4.5):

$$\begin{aligned} u(x_m,t_n) \approx u_m^n &= \sum_{j=0}^{n-1} \frac{x_m^j}{j!} u_x^{(j)}(0,t_n) \\ &+ \sum_{k=0}^{m} \frac{h_x^{\alpha}}{\Gamma(\alpha+2)} a_{m,k} \left(u_t^{(1)}(x_k,t_n) + f(x_k,t_n,u_k^n) \right) + \mathscr{O}\left(h_x^{2+\alpha}\right), \end{aligned} \tag{4.18}$$

where the coefficients $a_{m,k}$ are given in (4.12).

The temporal derivative $u_t^{(1)}(x_k,t_n)$ can be obtained by the upwind finite difference approximation in the t-direction:

$$u_t^{(1)}(x_k,t_n) = \frac{u_k^n - u_k^{n-1}}{h_t} + \mathscr{O}(h_t). \tag{4.19}$$

Therefore, it yields

$$\begin{aligned} u_m^n &= \sum_{j=0}^{n-1} \frac{x_m^j}{j!} u_x^{(j)}(0,t_n) \\ &+ \sum_{k=0}^{m} \frac{h_x^{\alpha}}{\Gamma(\alpha+2)} a_{m,k} \left(\frac{u_k^n - u_k^{n-1}}{h_t} + f(t_n,x_k,u_k^n) \right) + \mathscr{O}\left(h_x^{\alpha} h_t + h_x^{2+\alpha}\right). \end{aligned} \tag{4.20}$$

The explicit spline finite difference algorithm is obtained by neglecting $\mathscr{O}\left(h_x^{\alpha} h_t + h_x^{2+\alpha}\right)$:

$$u_m^n = \sum_{j=0}^{n-1} \frac{x_m^j}{j!} u_x^{(j)}(0, t_n) + \frac{h_x^{\alpha}}{\Gamma(\alpha+2)} \left(\frac{u_m^n - u_m^{n-1}}{h_t} + f(t_n, x_m, u_m^n) \right)$$

$$+ \sum_{k=0}^{m-1} \frac{h_x^{\alpha}}{\Gamma(\alpha+2)} a_{m,k} \left(\frac{u_k^n - u_k^{n-1}}{h_t} + f(t_n, x_k, u_k^n) \right). \tag{4.21}$$

We often employ an iteration procedure to obtain unknown variable u_m^n in (4.21) as the function f is usually nonlinear and u_m^n exists in the both sides of the equation (4.21). Moreover, we substitute a predicted value u_m^n into the right-hand side of (4.21) to achieve a better approximation. For this purpose, let ${}^p u_m^n$ be the predicted solution obtained by the Adams–Bashforth method [5] as

$${}^p u_m^n = \sum_{j=0}^{n-1} \frac{x_m^j}{j!} u_x^{(j)}(0, t_n)$$

$$+ \sum_{k=0}^{m-1} \frac{h_x^{\alpha}}{\Gamma(\alpha+1)} c_{m,k} \left(\frac{u_k^n - u_k^{n-1}}{h_t} + f(t_n, x_k, u_k^n) \right), \tag{4.22}$$

where

$$c_{m,k} = (m-k)^{\alpha} - (m-k-1)^{\alpha}. \tag{4.23}$$

Replacing ${}^p u_m^n$ in the right-hand side of (4.21) by (4.22) gives

$$u_m^n = \sum_{j=0}^{\lceil \alpha \rceil - 1} \frac{x_m^j}{j!} u_x^{(j)}(0, t_n) + \frac{h_x^{\alpha}}{\Gamma(\alpha+2)} \left(\frac{{}^p u_m^n - u_m^{n-1}}{h_t} + f(t_n, x_m, {}^p u_m^n) \right)$$

$$+ \sum_{k=0}^{m-1} \frac{h_x^{\alpha}}{\Gamma(\alpha+2)} a_{m,k} \left(\frac{u_k^n - u_k^{n-1}}{h_t} + f(t_n, x_k, u_k^n) \right). \tag{4.24}$$

Theorem 4.3.1 *Suppose that $Re(\alpha) > 0$ and consider the functions $u(x, t)$ and $u_x^{(\alpha)}(x, t) \in C^2(\Omega)$. If the function $f(t, x, u)$ holds the Lipschitz condition with respect to the second variable, then the error of the scheme (4.22)–(4.24) satisfies*

$$\|u(x_j, t_i) - u_j^i\|_{\infty} \leq C \left(h_x^{\alpha} h_t + h_x^{2+\alpha} \right), \tag{4.25}$$

where C is a constant independent of h_t and h_x.

Proof The inequality (4.25) holds for $j = i = 0$ as the initial and boundary conditions are given. Now, we shall prove that the inequality (4.25) also holds for $j = m$ and $i = n$, when it is held for $j = 0, 1, \ldots, m-1$ and $i = 0, 1, \ldots, n-1$.

Firstly, we observe the error of predicted value ${}^{p}u_m^n$. Subtracting (4.22) from (4.17) yields

$$\left|u(x_m, t_n) - {}^{p}u_m^n\right| = \left|\left[\mathscr{I}_{0,x}^{\alpha}\left(u_t^{(1)}(x,t) + f(t,x,u(x,t))\right)\right]_{(x_m,t_n)} - \sum_{k=0}^{m-1} \frac{h_x^{\alpha}}{\Gamma(\alpha+1)} c_{m,k}\left(u_t^{(1)}(x_k,t_n) + f(t_n, x_k, u_k^n)\right)\right|.$$

Using (4.13), we have

$$\begin{aligned}\left|u(x_m, t_n) - {}^{p}u_m^n\right| \le &\left|\left[\mathscr{I}_{0,x}^{\alpha} u_x^{(\alpha)}(x,t)\right]_{(x_m,t_n)} - \sum_{k=0}^{m-1} \frac{h_x^{\alpha}}{\Gamma(\alpha+1)} c_{m,k}[u_x^{(\alpha)}(x,t)]_{(x_m,t_n)}\right| \\ &+ \sum_{k=0}^{m-1} \frac{h_x^{\alpha}}{\Gamma(\alpha+1)} c_{m,k}\left(\left|u_t^{(1)}(x_k,t_n) - (u_t^{(1)})_k^n\right|\right. \\ &\left. + \left|f(t_n, x_k, u(x_k,t_n)) - f(t_n, x_k, u_k^n)\right|\right),\end{aligned}$$

and by applying the Lipschitz condition (4.15) and $\sum_{k=0}^{m-1} c_{m,k} = m^{\alpha}$, we get

$$\left|u(x_m,t_n) - {}^{p}u_m^n\right| \le c_1 m^{\alpha} h_x^{2+\alpha} + c_2 h_t h_x^{\alpha} + c_3 l_f \left|u(x_k,t_n) - u_k^n\right| \le C\left(h_x^{2+\alpha} + h_t h_x^{\alpha}\right).$$

Then, we obtain the error of the corrected value.
From (4.17) and (4.24), we finally have

$$\begin{aligned}|u(x_m,t_n) - u_m^n| = &\left|\left[\mathscr{I}_{0,x}^{\alpha}\left(u_t^{(1)}(x,t) + f(t,x,u(x,t))\right)\right]_{(x_m,t_n)}\right. \\ &- \frac{h_x^{\alpha}}{\Gamma(\alpha+2)}\left(\frac{{}^{p}u_m^n - u_m^{n-1}}{h_t} + f(t_n, x_m, {}^{p}u_m^n)\right) \\ &\left.- \sum_{k=0}^{m-1} \frac{h_x^{\alpha}}{\Gamma(\alpha+2)} a_{m,k}\left(\frac{u_k^n - u_k^{n-1}}{h_t} + f(t_n, x_k, u_k^n)\right)\right| \\ \le &\left|[\mathscr{I}_{0,x}^{\alpha} u_x^{(\alpha)}(x,t)]_{(x_m,t_n)} - \sum_{k=0}^{m-1} \frac{h_x^{\alpha}}{\Gamma(\alpha+2)} a_{m,k}[u_x^{(\alpha)}(x,t)]_{(x_m,t_n)}\right|\end{aligned}$$

$$+\sum_{k=0}^{m-1}\frac{h_x^{\alpha}}{\Gamma(\alpha+2)}a_{m,k}\Big[|u_t^{(1)}(x_k,t_n)-(u_t^{(1)})_k^n|$$
$$+|f(t_n,x_k,u(x_k,t_n))-f(t_n,x_k,u_k^n)|\Big]$$
$$+\frac{h_x^{\alpha}}{\Gamma(\alpha+2)}\Big[|u_t^{(1)}(x_m,t_n)-{}^{p}(u_t^{(1)})_m^n|+|f(t_n,x_m,u(x_m,t_n))$$
$$-f(t_n,x_m,{}^{p}u_m^n)|\Big]$$
$$\leq\ c_4m^{\alpha}h_x^{2+\alpha}+c_5h_th_x^{\alpha}+c_6m^{\alpha}l_fh_x^{2+\alpha}+c_7h_th_x^{\alpha}+c_8m^{\alpha}l_fh_x^{2+\alpha}$$
$$\leq\ C(h_x^{2+\alpha}+h_th_x^{\alpha}).$$

Theorem 4.3.2 *Let u_m^n and w_m^n be the numerical solutions of (4.21) at the point (x_m,t_n) and that the boundary conditions are given by $u_x^{(j)}(0,t_n)$ and $w_0^{(j)}(t_n)$, respectively. In addition, assume that*

$$|u_j^i-w_j^i|\leq\kappa\|u_0-w_0\|_{\infty} \tag{4.26}$$

for $j=0,1,\ldots,m-1$ and $i=0,1,\ldots,n-1$.

If there exists positive κ, independent of h_t and h, such that

$$\|u_m^n-w_m^n\|_{\infty}\leq\kappa\|u_0-w_0\|_{\infty}, \tag{4.27}$$

for any m and n, then the new scheme (4.22)–(4.24) is unconditionally stable.

Proof We prove the inequality also holds for $j=m$ and $i=n$.
According to the expression of (4.20), we get

$$|u_m^n-w_m^n|=\Bigg|\sum_{k=0}^{\lceil\alpha\rceil-1}\frac{x_m^k}{k!}u_0^{(k)}(t_n)+\frac{h_x^{\alpha}}{\Gamma(\alpha+2)}\left(\frac{{}^{p}u_m^n-u_m^{n-1}}{h_t}+f(t_n,x_m,{}^{p}u_m^n)\right)$$
$$+\sum_{k=0}^{m-1}\frac{h_x^{\alpha}}{\Gamma(\alpha+2)}a_{m,k}\left(\frac{u_k^n-u_k^{n-1}}{h_t}+f(t_n,x_k,u_k^n)\right)$$
$$-\sum_{k=0}^{\lceil\alpha\rceil-1}\frac{x_m^k}{k!}w_0^{(k)}(t_n)+\frac{h_x^{\alpha}}{\Gamma(\alpha+2)}\left(\frac{{}^{p}w_m^n-w_m^{n-1}}{h_t}-f(t_n,x_m,{}^{p}w_m^n)\right)$$
$$+\sum_{k=0}^{m-1}\frac{h_x^{\alpha}}{\Gamma(\alpha+2)}a_{m,k}\left(\frac{w_k^n-w_k^{n-1}}{h_t}+f(t_n,x_k,w_k^n)\right)\Bigg|$$
$$\leq\sum_{k=0}^{\lceil\alpha\rceil-1}\frac{x_m^k}{k!}\left|u_0^{(k)}(t_n)-w_0^{(k)}(t_n)\right|$$

$$+\sum_{k=0}^{m-1} \frac{h_x^\alpha}{\Gamma(\alpha+2)} a_{m,k} \left(\left| \frac{u_k^n - u_k^{n-1}}{h_t} - \frac{w_k^n - w_k^{n-1}}{h_t} \right| + \left| f(t_n, x_k, u_k^n) \right. \right.$$

$$\left. \left. - f(t_n, x_k, w_k^n) \right| \right)$$

$$+\frac{h_x^\alpha}{\Gamma(\alpha+2)} \left(\left| \frac{{}^p u_m^n - u_m^{n-1}}{h_t} - \frac{{}^p w_m^n - w_m^{n-1}}{h_t} \right| + \left| f(t_n, x_m, {}^p u_m^n) \right. \right.$$

$$\left. \left. - f(t_n, x_m, {}^p w_m^n) \right| \right)$$

$$\leq \kappa_1 \|u_0 - w_0\|_\infty + \kappa_2 \left(\max_{0<k<m-1} |u_k^n - w_k^n| \right.$$

$$\left. + \max_{0<k<m-1} |u_k^{n-1} - w_k^{n-1}|) + l_f \max_{0<k<m-1} |u_k^n - w_k^n| \right)$$

$$+\kappa_3 \left(\max_{0<k<m-1} |{}^p u_k^n - {}^p w_k^n| + \max_{0<k<m-1} |u_k^{n-1} - w_k^{n-1}|) \right.$$

$$\left. + l_f \max_{0<k<m-1} |{}^p u_k^n -{}^p w_k^n| \right)$$

and, therefore, using (4.26), we obtain

$$|u_m^n - w_m^n| \leq \kappa \|u_0 - w_0\|_\infty.$$

This expression means that proposed scheme is unconditionally stable with respect to the initial conditions.

4.3.2 Time-FPDE

Consider the following time-FPDE with the solution $u(x,t) \in C^2(\Omega)$, $\Omega = (0, L) \times (0, T) \subseteq \mathbb{R}^2$,

$$u_t^{(\alpha)}(x,t) = u_x^{(2)}(x,t) + f(t, x, u(x,t)), \quad 0 < \alpha \leq 1, \tag{4.28}$$

subject to the initial condition

$$u(x, 0) = g(x), \qquad x \in (0, L), \tag{4.29}$$

and boundary conditions

$$u(0, t) = h(t), \qquad u(L, t) = q(t), \qquad t \in (0, T), \tag{4.30}$$

where $g(x)$, $h(t)$, and $q(t)$ are assumed to be smooth functions and the nonlinear source term $f(x, t, u)$ satisfies the Lipschitz condition (4.15).

Using the finite difference approximation, the second-order derivative $u_x^{(2)}(x, t)$ is approximated at (x_m, t_n) as

$$u_x^{(2)}(x_m, t_n) \approx \frac{u(x_{m+1}, t_n) - 2u(x_m, t_n) + u(x_{m-1}, t_n)}{h_x^2} + \mathscr{O}(h_x^2). \tag{4.31}$$

The approximate solution of (4.28)–(4.30) is obtained by discretizing the time-fractional derivative using Proposition 4.2.2 as

$$\sum_{k=0}^{n} \frac{h_t^{1-\alpha}}{\Gamma(3-\alpha)} b_{k,n} u_t^{(1)}(x_m, t_n) = \frac{u(x_{m+1}, t_n) - 2u(x_m, t_n) + u(x_{m-1}, t_n)}{h_x^2} + f(x_m, t_n, u(x_m, t_n)) + \mathscr{O}\left(h_t^2 + h_x^2 + \left(\frac{h_t}{h_x}\right)^2\right). \tag{4.32}$$

Following the idea of the Du Fort–Frankel scheme [6], u_m^n is replaced by its average in time $\dfrac{u_m^{n+1} + u_m^{n-1}}{2}$ resulting in

$$\sum_{k=0}^{n} \frac{h_t^{1-\alpha_k} b_{k,n}}{\Gamma(3-\alpha_k)} \left(\frac{u_m^{k-1} - u_m^{k+1}}{2h_t}\right) = \frac{u_{m+1}^n - u_m^{n+1} - u_m^{n-1} + u_{m-1}^n}{h_x^2} + f(x_m, t_n, u_m^n). \tag{4.33}$$

Rearranging the terms in (4.33), the three-level explicit spline finite difference method is obtained as

$$\begin{aligned} u_m^{n+1} &= \frac{2\vartheta}{w_n + 2\vartheta}(u_{m+1}^n + u_{m-1}^n) + \frac{w_n - 2\vartheta}{w_n + 2\vartheta} u_m^{n-1} \\ &\quad + \sum_{k=0}^{n-1} \frac{w_{k,n}}{w_n + 2\vartheta}(u_m^{k-1} - u_m^{k+1}) + \frac{2h_t}{w_n + 2\vartheta} f(x_m, t_n, u_m^n), \end{aligned} \tag{4.34}$$

where $w_{k,n} = \dfrac{h_t^{1-\alpha} b_{k,n}}{\Gamma(3-\alpha)}$, $w_n = \dfrac{h_t^{1-\alpha}}{\Gamma(3-\alpha)}$ and $\vartheta = \dfrac{h_t}{h_x^2}$.

This scheme is conditionally consistent, because the truncation error term $\mathscr{O}\left(\frac{h_t^2}{h_x^2}\right)$ in the local truncation error $\mathscr{O}\left(h_t^2 + h_x^2 + \left(\frac{h_t}{h_x}\right)^2\right)$ of (4.32) leads to the consistency condition when $\Delta x \to 0$, i.e., $\frac{\Delta t}{\Delta x} \to 0$.

Let the round-off error be

$$\mathscr{E}_m^n = U_m^n - u_m^n, \tag{4.35}$$

where U_m^n is an approximate solution of (4.34) for $n = 0, 1, \ldots, N$ and $m = 0, 1, \ldots, M$. Then, the round-off error satisfies the following equation using the discretized (4.34):

$$\mathscr{E}_m^{n+1} = \frac{2\vartheta}{w_n+2\vartheta}(\mathscr{E}_{m+1}^n+\mathscr{E}_{m-1}^n)+\frac{w_n-2\vartheta}{w_n+2\vartheta}\mathscr{E}_m^{n-1}+\sum_{k=0}^{n-1}\frac{w_{k,n}}{w_n+2\vartheta}(\mathscr{E}_m^{k-1}-\mathscr{E}_m^{k+1})$$
$$+\frac{2h_t}{w_n+2\vartheta}(f(x_m,t_n,U_m^n)-f(x_m,t_n,u_m^n)), \tag{4.36}$$

such that

$$\mathscr{E}_0^n = \mathscr{E}_M^n = 0, \qquad n = 0, 1, \ldots, N. \tag{4.37}$$

Theorem 4.3.3 *The explicit approximation (4.34) with the initial and boundary conditions*

$$\begin{cases} u_m^0 = g(mh_x), & m = 0, 1, \ldots, M \\ u_0^n = h(nh_t), & u_M^n = q(nh_t),\ n = 0, 1, \ldots, N \end{cases} \tag{4.38}$$

is unconditionally stable and

$$\|\mathscr{E}^n\|_\infty \le \beta\|\mathscr{E}^0\|_\infty, \tag{4.39}$$

for $n = 0, 1, \ldots, N$, where $\|\mathscr{E}^n\|_\infty = \max\limits_{0\le m\le M}|\mathscr{E}_m^n|$, and β is a positive number independent of n, h_x, and h_t.

Proof The inequality (4.39) holds for $n = 0$ as the initial and boundary conditions are feasible. We shall prove that the inequality (4.39) holds for $n = r + 1$ as it is assumed to be held for $n = 0, 1, \ldots, r$. Using (4.36) for $n = 0, 1, \ldots, r$ and $m = 0, 1, \ldots, M$, we can write

$$\begin{aligned} |\mathscr{E}_m^{r+1}| \le\ & \frac{2\vartheta}{w_r+2\vartheta}(|\mathscr{E}_{m+1}^r|+|\mathscr{E}_{m-1}^r|)+\frac{w_r-2\vartheta}{w_r+2\vartheta}|\mathscr{E}_m^{r-1}| \\ & +\sum_{k=0}^{r-1}\frac{w_{k,r}}{w_r+2\vartheta}(|\mathscr{E}_m^{k-1}|+|\mathscr{E}_m^{k+1}|)+\frac{2l_f h_t}{w_r+2\vartheta}|\mathscr{E}_m^r| \\ \le\ & \frac{2\vartheta}{w_r+2\vartheta}\|\mathscr{E}^r\|_\infty+\frac{w_r-2\vartheta}{w_r+2\vartheta}\|\mathscr{E}^{r-1}\|_\infty \\ & +\sum_{k=0}^{r-1}\frac{2w_{k,r}}{w_r+2\vartheta}(\|\mathscr{E}^{k-1}\|_\infty)+\frac{2l_f h_t}{w_r+2\vartheta}\|\mathscr{E}^r\|_\infty, \end{aligned} \tag{4.40}$$

where l is the Lipschitz constant. Considering $\sum_{k=0}^{r-1} w_{k,r} < 1$ [19], it results

$$|\mathscr{E}_m^{r+1}| \leq \left(\frac{4\vartheta}{w_r + 2\vartheta} + \frac{w_r - 2\vartheta}{w_r + 2\vartheta} + \frac{2l_f h_t}{w_r + 2\vartheta}\right) \|\mathscr{E}^r\|_\infty + \beta_1 \|\mathscr{E}^0\|_\infty$$
$$\leq \left(1 + \frac{2l_f h_t}{w_r + 2\vartheta}\right) \|\mathscr{E}^r\|_\infty + \beta_1 \|\mathscr{E}^0\|_\infty \leq (1 + l_f h_t)\|\mathscr{E}^r\|_\infty + \beta_1 \|\mathscr{E}^0\|_\infty, \quad (4.41)$$

or

$$\|\mathscr{E}^{r+1}\|_\infty \leq \left(1 + l_f h_t\right)^{r+1} \|\mathscr{E}^0\|_\infty + \beta_1 \|\mathscr{E}^0\|_\infty \leq \mathrm{e}^{(r+1)l_f h_t} \|\mathscr{E}^0\|_\infty + \beta_1 \|\mathscr{E}^0\|_\infty$$
$$\leq \mathrm{e}^{t_{r+1} l_f} \|\mathscr{E}^0\|_\infty + \beta_1 \|\mathscr{E}^0\|_\infty \leq \mathrm{e}^{T l_f} \|\mathscr{E}^0\|_\infty + \beta_1 \|\mathscr{E}^0\|_\infty$$
$$= (\beta_1 + \beta_2)\|\mathscr{E}^0\|_\infty = \beta \|\mathscr{E}^0\|_\infty,$$

where $\beta = \beta_1 + \beta_2$. Therefore, for any arbitrary initial rounding error $\mathscr{E}^0$, we have a positive β independent from n, h_t, and h_x, such that

$$\|\mathscr{E}^n\|_\infty \leq \beta \|\mathscr{E}^0\|_\infty. \quad (4.42)$$

This completes the proof and shows that the proposed scheme is stable.

4.4 Applications

In this section, the stability and efficiency of the proposed method are illustrated in four special cases of the nonlinear space and time-fractional KPPE including the Newell–Whitehead, FitzHugh–Nagumo, Burgers–Huxley, and Chafee–Infante equations. The algorithms are encoded in Maple. The fractional KPPE is defined as

$$u_x^{(\alpha)}(x,t) = u_t^{(\beta)}(x,t) + au^3(x,t) + bu^2(x,t) + cu(x,t) + g(x,t), \quad (4.43)$$

where $1 < \alpha \leq 2$, $0 < \beta \leq 1$ and $a, b, c \in \mathbb{R}$ are constants.

We analyze the computational accuracy in the perspective of the maximum of the absolute error (MAE). If the exact solution is available, then the MAE is defined as

$$\|\mathscr{E}_{M,N}\|_\infty = \max_{1 \leq n \leq N-1, 1 \leq m \leq M-1} | u(x_m, t_n) - {}_M^N u_m^n |, \quad (4.44)$$

else it is defined as

$$\|\mathscr{E}_{M,N}\|_\infty^\dagger = \max_{1 \leq n \leq N-1, 1 \leq m \leq M-1} | {}_{2M}^{2N} u_m^n - {}_M^N u_m^n |, \quad (4.45)$$

where ${}^{N}_{M}u^{n}_{m}$ is the approximate solution of $u(x,t)$ at (x_m, t_n) with N and M numbers of interior time and space mesh points, respectively. Furthermore, the experimental convergence order (ECO) is estimated by

$$p_{(\cdot)} = \log_{\frac{h_1(\cdot)}{h_2(\cdot)}} \frac{\|\mathscr{E}_1\|_\infty}{\|\mathscr{E}_2\|_\infty}, \tag{4.46}$$

where $\|\mathscr{E}_i\|_\infty$ denotes the MAE for the step $h_i(\cdot)$, $i = \{1, 2\}$.

Example 4.4.1 The space-fractional Newell–Whitehead–Segel equation has been applied to different problems such as Rayleigh–Benard convection, Faraday instability, nonlinear optics, chemical reactions, and biological systems [13]. It is given by (4.43) when $\beta = 1$, $a = 1$, $b = 0$, $c = -1$, and

$$g(x,t) = \frac{e^t}{\Gamma(3-\alpha)}\left(2x^{2-\alpha} + x^3 e^{2t}\Gamma(3-\alpha)(1-x)^3\right). \tag{4.47}$$

Let us choose the initial and boundary conditions as

$$\begin{cases} u(x,0) = x^2 - x, & 0 \le x \le 1 \\ u(0,t) = 0, & 0 \le t \le 1 \\ u_x^{(1)}(0,t) = -e^t, & 0 \le t \le 1 \end{cases}. \tag{4.48}$$

The exact solution is $u(x,t) = e^t\left(x^2 - x\right)$.

Figure 4.1a and b shows the numerical solution of Example 4.4.1 obtained by the proposed method when $\alpha = 1.75$ and $h_x = h_t = \frac{1}{32}$. Figure 4.1a shows the obtained solution in the temporal and spatial domains. Figure 4.1b shows the exact solution along the approximated solution for different values of $t = \{0.25, 0.5, 0.75, 1\}$. Table 4.1 shows the MAEs and ECOs for different values of α and step sizes in order to analyze the performance of the proposed scheme. The results show that increasing α increases both the MAEs and ECOs, but they are reduced when increasing the step sizes.

Example 4.4.2 The space-fractional FitzHugh–Nagumo equation is a reaction–diffusion equation describing the propagation of electrical signals in nerve axons and other biological tissues [10, 11, 22]. It is given by (4.43) when $\beta = 1$, $a = 1$, $b = -3$, $c = 2$, and

$$g(x,t) = e^{xt}\left(erf(\sqrt{xt})t^{\frac{3}{2}} - e^{2xt} + 3e^{xt} - x - 2\right). \tag{4.49}$$

Let us choose the initial and boundary conditions as

$$\begin{cases} u(x,0) = 1, & 0 \le x \le 1 \\ u(0,t) = 1, & 0 \le t \le 1 \\ u_x^{(1)}(0,t) = t, & 0 \le t \le 1 \end{cases}, \tag{4.50}$$

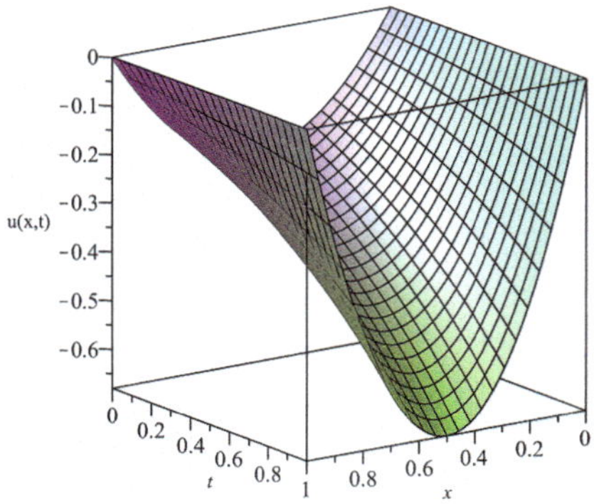

(a)

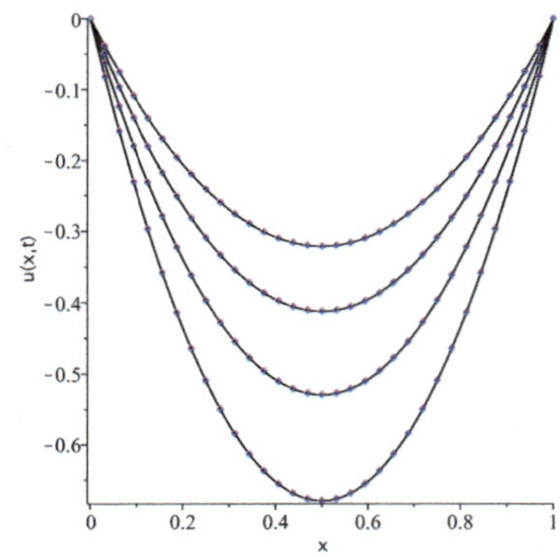

(b)

Fig. 4.1 The numerical solution of Example 4.4.1 with step size $h_x = h_t = \frac{1}{32}$ and $\alpha = 1.75$. (**a**) The numerical solution of $u(x, t)$. (**b**) The exact (blue line) and approximated (black circle) solutions of $u(x, t)$ when $t = \{0.25, 0.5, 0.75, 1\}$

Table 4.1 Comparison of the MAEs and ECOs of Example 4.4.1 for different values of α and step sizes h_x and h_t

	$\alpha = 1.25$		$\alpha = 1.50$		$\alpha = 1.75$	
$h_x = h_t$	$\Vert\mathscr{E}\Vert_\infty$	$p_x = p_t$	$\Vert\mathscr{E}\Vert_\infty$	$p_x = p_t$	$\Vert\mathscr{E}\Vert_\infty$	$p_x = p_t$
$\frac{1}{32}$	9.05×10^{-4}	–	3.57×10^{-3}	–	1.09×10^{-2}	–
$\frac{1}{64}$	2.76×10^{-4}	1.713	1.26×10^{-3}	1.503	4.50×10^{-3}	1.276
$\frac{1}{128}$	8.46×10^{-5}	1.706	4.48×10^{-4}	1.491	1.88×10^{-3}	1.259

and hence the exact solution is $u(x, t) = e^{xt}$ for $\alpha = \frac{3}{2}$.

Figure 4.2 depicts the exact and approximate solutions of Example (4.4.2) for various values of $t = \{0.25, 0.5, 0.75, 1\}$ when $h_x = h_t = \frac{1}{32}$ and $\alpha = \frac{3}{2}$. Similar to the previous example, Table 4.2 shows that increasing α increases both the MAEs and ECOs, but they are reduced by increasing the step sizes. It is also shown that the obtained numerical approximations are in a good agreement with the analytical solutions.

Example 4.4.3 Consider a time-fractional diffusion equation as (4.43) when $\alpha = 2$, $a = b = c = 0$ and $g(x, t) = -\frac{3\sqrt{\pi}}{4\Gamma(2.5-\alpha)} t^{1.5-\beta} x^4 (x - 1)$. The analytical solution is nonsmooth at $t = 0$ and is given by $u(x, t) = x^4(x - 1)t^{1.5}$ when the initial and boundary conditions are [9, 20].

$$\begin{cases} u(x, 0) = 0, & 0 \leq x \leq 1 \\ u(0, t) = u(1, t) = 0, & 0 \leq t \leq 1 \end{cases}. \tag{4.51}$$

Let the convergence orders in time and space be denoted by p_t and p_x, respectively, so that we have $p_x = 2p_t$ for $h_t = h_x^2$. The numerical solution

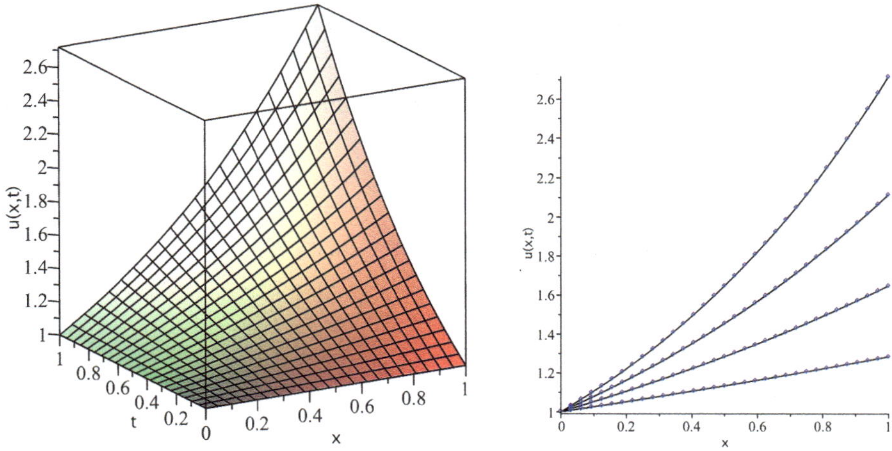

(a) (b)

Fig. 4.2 The approximate solution of Example 4.4.2 when $h_x = h_t = \frac{1}{32}$ and $\alpha = \frac{3}{2}$. (**a**) The plot of $u(x,t)$ versus (x,t). (**b**) The exact (blue line) and approximated (black circle) solutions of $u(x,t)$ for $t = \{0.25, 0.5, 0.75, 1\}$

Table 4.2 Comparison of the MAEs and ECOs of Example 4.4.2 for different values of α, h_x, and h_t

	$\alpha = 1.25$		$\alpha = 1.50$		$\alpha = 1.75$	
$h_x = h_t$	$\\|\mathscr{E}\\|_\infty^\dagger$	$p_x = p_t$	$\\|\mathscr{E}\\|_\infty$	$p_x = p_t$	$\\|\mathscr{E}\\|_\infty^\dagger$	$p_x = p_t$
$\frac{1}{32}$	1.42×10^{-3}	–	2.81×10^{-2}	–	6.78×10^{-4}	–
$\frac{1}{64}$	4.49×10^{-4}	1.661	9.86×10^{-3}	1.512	3.44×10^{-4}	0.979
$\frac{1}{128}$	1.43×10^{-4}	1.650	3.48×10^{-3}	1.501	1.76×10^{-4}	0.966

of Example 4.4.3 is obtained for different values of β and step sizes. Figure 4.3a shows the approximation solution of Example 4.4.3 when $\beta = 0.75$ and $h_x = \frac{1}{16}$. Figure 4.3b shows the exact and numerical solutions for different values of $t = \{0.25, 0.5, 0.75, 1\}$. In addition, Table 4.3 lists the MAEs and ECOs for different values of α and step sizes, revealing that the obtained approximations are in a good agreement with the analytical solutions.

Example 4.4.4 The FitzHugh–Nagumo equation is a reaction–diffusion equation describing the propagation of electrical signals in nerve axons and other biological tissues [10, 11, 20, 22]. This equation is given by (4.43) when $\alpha = 2$, $a = 1$, $b = -\frac{5}{2}$, $c = \frac{2}{3}$, and $g(x,t) = 0$. The exact solution is $u(x,t) = \left(1 + e^{\frac{-\sqrt{2}}{2}\left(x + \frac{3\sqrt{2}}{2}t\right)}\right)^{-1}$ when $\beta = 1$ and the initial and boundary conditions are

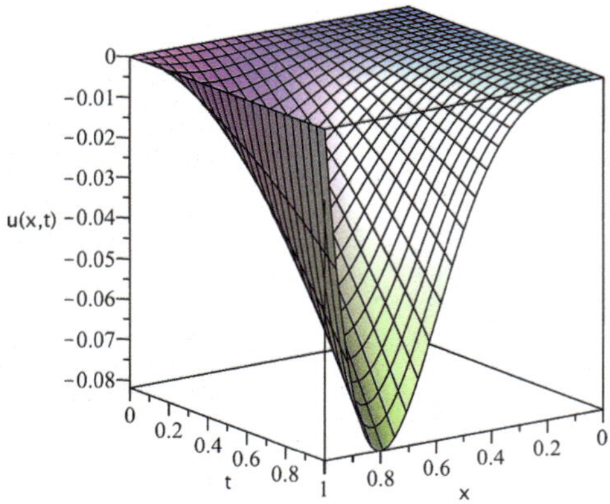

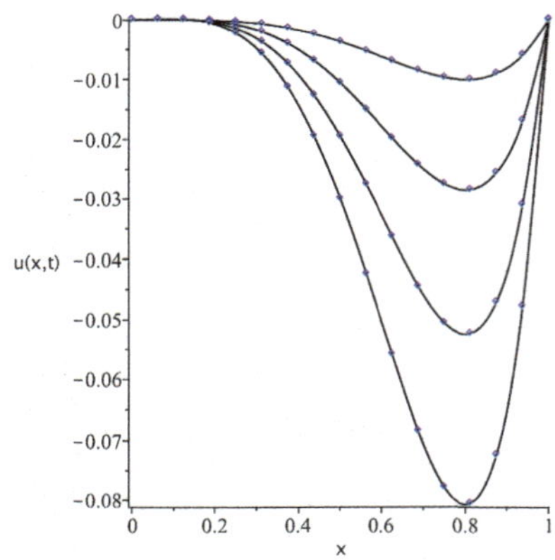

Fig. 4.3 Solution of example 4.4.3 with $h_x = \frac{1}{16}$ and $h_t = h_x^2$ and $\beta = 0.75$. (**a**) Plot of the approximate solution. (**b**) Exact solution (line) and approximate solution (points)

Table 4.3 Comparison of MAEs and ECOs of Example 4.4.3 for different values of β and time step sizes h_x when $p_x = 2p_t$

	$\beta = 0.25$		$\beta = 0.50$		$\beta = 0.75$	
h_x	$\Vert\mathscr{E}_{h_t=h_x^2}\Vert_\infty^\dagger$	p_x	$\Vert\mathscr{E}_{h_t=h_x^2}\Vert_\infty^\dagger$	p_x	$\Vert\mathscr{E}_{h_t=h_x^2}\Vert_\infty^\dagger$	p_x
$\frac{1}{8}$	3.59×10^{-3}	–	3.44×10^{-3}	–	3.40×10^{-3}	–
$\frac{1}{16}$	8.82×10^{-4}	2.025	8.74×10^{-4}	1.977	8.62×10^{-4}	1.980
$\frac{1}{32}$	2.23×10^{-4}	1.984	2.19×10^{-4}	1.997	2.16×10^{-4}	1.997

$$\begin{cases} u(x,0) = \left(1 + e^{0.5\sqrt{2}x}\right)^{-1}, & 0 \le x \le 1 \\ u(0,t) = \left(1 + e^{\frac{1}{6}t}\right)^{-1}, & 0 \le t \le 1 \\ u(1,t) = \left(1 + e^{\frac{\sqrt{2}}{2}L + \frac{1}{6}t}\right)^{-1}, & 0 \le t \le 1 \end{cases}. \tag{4.52}$$

Figure 4.4a illustrates the numerical solution by means of the proposed method when $\beta = 0.75$, and Fig. 4.4b shows the numerical solution for different values of $t = \{0.25, 0.5, 0.75, 1\}$ when $h_x = \frac{1}{16}$. Similar to the previous examples, Table 4.4 includes the MAEs and ECOs of Example 4.4.4.

4.5 Conclusion

In this chapter, two stable numerical methods were introduced to solve nonlinear space and time-fractional partial differential equations. The discretization of the time-fractional integral and the space-fractional derivative were accomplished using

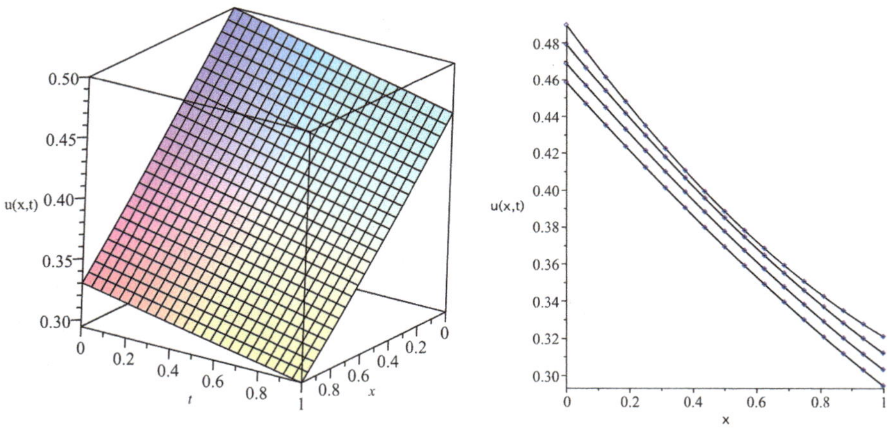

(a) (b)

Fig. 4.4 The solution of Example 4.4.4 when $h_x = \frac{1}{16}$, $h_t = h_x^2$, and $\beta = 0.75$. (**a**) Plot of the approximate solution. (**b**) The approximate solution for $t = \{0.25, 0.5, 0.75, 1\}$

Table 4.4 Comparison of the MAEs and ECOs of Example 4.4.4 for different values of β and h_x when $p_x = 2p_t$

	$\beta = 0.25$		$\beta = 0.50$		$\beta = 0.75$	
h_x	$\Vert\mathscr{E}_{h_t=h_x^2}\Vert_\infty^\dagger$	p_x	$\Vert\mathscr{E}_{h_t=h_x^2}\Vert_\infty^\dagger$	p_x	$\Vert\mathscr{E}_{h_t=h_x^2}\Vert_\infty^\dagger$	p_x
$\frac{1}{8}$	8.29×10^{-3}	–	2.41×10^{-4}	–	1.13×10^{-5}	–
$\frac{1}{16}$	1.87×10^{-3}	2.148	5.23×10^{-5}	2.204	2.65×10^{-6}	2.092
$\frac{1}{32}$	0.39×10^{-3}	2.261	7.39×10^{-6}	2.823	6.05×10^{-7}	2.130

a linear B-spline approximation. It was shown that the proposed methods are unconditionally stable. The advantages of the proposed algorithms were shown in four practical examples. It was shown that the suggested schemes are computationally efficient and the obtained solutions converge to the analytic solution.

References

1. Carella, A.R., Dorao, C.A.: Least-Squares spectral method for the solution of a fractional advection–dispersion equation. J. Comput. Phys. **232**(1), 33–45 (2013). Doi: https://doi.org/10.1016/j.jcp.2012.04.050
2. Chen, C.M., Liu, F., Anh, V., Turner, I.: Numerical schemes with high spatial accuracy for a variable-order anomalous subdiffusion equation. SIAM J. Sci. Comput. **32**(4), 1740–1760 (2010). Doi: https://doi.org/10.1137/090771715
3. Chen, W., Zhang, J., Zhang, J.: A variable-order time-fractional derivative model for chloride ions sub-diffusion in concrete structures. Fractional Calc. Appl. Anal. 16(1), 76–92 (2013). Doi: https://doi.org/10.2478/s13540-013-0006-y
4. Diethelm, K.: *The Analysis of Fractional Differential Equations*. Lecture Notes in Mathematics (2010). Doi: https://doi.org/10.1007/978-3-642-14574-2

5. Diethelm, K., Ford, N.J., Freed, A.D.: A predictor-corrector approach for the numerical solution of fractional differential equations. Nonlinear Dynamics **29**(1/4), 3–22 (2002). Doi: https://doi.org/10.1023/a:1016592219341
6. Du Fort, E.C., Frankel, S.P.: *Conditions in the Numerical Treatment of Parabolic Differential Equations*. Mathematical Tables and Other Aids to Computation, Vol. 7(43), pp. 135–152 (1953)
7. Ervin, V.J., Roop, J.P.: Variational formulation for the stationary fractional advection dispersion equation. Numer. Methods Partial Differ. Equ. **22**(3), 558–576 (2006). Doi: https://doi.org/10.1002/num.20112
8. Ervin, V.J., Heuer, N., Roop, J.P.: Numerical approximation of a time dependent, nonlinear, space-fractional diffusion equation. SIAM J. Numer. Anal. **45**(2), 572–591 (2007). Doi: https://doi.org/10.1137/050642757
9. Ferrás, L.L., Ford, N.J., Morgado, M.L., Rebelo, M.: *A numerical method for the solution of the time-fractional diffusion equation*. In: International Conference on Computational Science and Its Applications. Springer, Cham (2014). Doi: https://doi.org/10.1007/978-3-319-09144-0-9
10. FitzHugh, R.: Mathematical models of threshold phenomena in the nerve membrane. Bull. Math. Biophys. **17**(4), 257–278 (1955). Doi: https://doi.org/10.1007/bf02477753
11. FitzHugh, R.: Impulses and physiological states in theoretical models of nerve membrane. Biophys. J. **1**(6), 445–466 (1961). Doi: https://doi.org/10.1016/s0006-3495(61)86902-6
12. Fu, Z.J., Chen, W., Ling, L.: Method of approximate particular solutions for constant- and variable-order fractional diffusion models. Eng. Anal. Boundary Elem. **57**, 37–46 (2015). Doi: https://doi.org/10.1016/j.enganabound.2014.09.003
13. Jassim, H.K.: Homotopy perturbation algorithm using Laplace transform for Newell–Whitehead–Segel equation. Int. J. Adv. Appl. Math. Mech. **2**(4), 8–12 (2015)
14. Li, C., Zeng, F.: *Numerical Methods for Fractional Calculus*. Chapman and Hall/CRC (2015)
15. Li, C., Yi, Q., Chen, A.: Finite difference methods with non-uniform meshes for nonlinear fractional differential equations. J. Comput. Phys. **316**, 614–631 (2016). Doi: https://doi.org/10.1016/j.jcp.2016.04.039
16. Lin, R., Liu, F., Anh, V., Turner, I.: Stability and convergence of a new explicit finite-difference approximation for the variable-order nonlinear fractional diffusion equation. Appl. Math. Comput. **212**(2), 435–445 (2009). Doi: https://doi.org/10.1016/j.amc.2009.02.047
17. Liu, F., Zhuang, P., Turner, I., Anh, V., Burrage, K.: A semi-alternating direction method for a 2-D fractional FitzHugh-Nagumo monodomain model on an approximate irregular domain. J. Comput. Phys. **293**, 252–263 (2015). Doi: https://doi.org/10.1016/j.jcp.2014.06.001
18. Miller, K.S., Ross, B.: *An Introduction to the Fractional Calculus and Fractional Differential Equations*. Wiley, New York (1993)
19. Moghaddam, B.P., Tenreiro Machado, J.A., Morgado, M.L.: Numerical approach for a class of distributed order time fractional partial differential equations. Appl. Numer. Math. **136**, 152–162 (2019). Doi: https://doi.org/10.1016/j.apnum.2018.09.019
20. Moghaddam, B.P., Tenreiro Machado, J.A.: A stable three-level explicit spline finite difference scheme for a class of nonlinear time variable order fractional partial differential equations. Comput. Math. Appl. **73**(6), 1262–1269 (2017). Doi: https://doi.org/10.1016/j.camwa.2016.07.010
21. Moghaddam, B.P., Yaghoobi, S., Tenreiro Machado, J.A.: An extended predictor–corrector algorithm for variable-order fractional delay differential equations. J. Comput. Nonlinear Dyn. **11**(6), 061001 (2016). Doi: https://doi.org/10.1115/1.4032574
22. Nagumo, J., Arimoto, S., Yoshizawa, S.: An active pulse transmission line simulating nerve axon. Proc. IRE **50**(10), 2061–2070 (1962). Doi: https://doi.org/10.1109/jrproc.1962.288235
23. Podlubny, I.: *Fractional Differential Equations: An Introduction to Fractional Derivatives, Fractional Differential Equations, to Methods of Their Solution and Some of Their Applications*. Academic Press, Elsevier, San Diego (1998)

24. Roop, J.P.: Computational aspects of FEM approximation of fractional advection dispersion equations on bounded domains in $\mathbb{R}^2$. J. Comput. Appl. Math. **193**(1), 243–268 (2006). Doi: https://doi.org/10.1016/j.cam.2005.06.005
25. Samko, S.G., Kilbas, A.A., Marichev, O.I.: *Fractional Integrals and Derivatives: Theory and Applications*. Gordon and Breach, London (1993)
26. Sun, H., Chen, W., Chen, Y.: Variable-order fractional differential operators in anomalous diffusion modeling. Phys. A Stat. Mech. Appl. **388**(21), 4586–4592 (2009). Doi: https://doi.org/10.1016/j.physa.2009.07.024
27. West, B.J., Bologna, M., Grigolini, P.: *Physics of Fractal Operators. Institute for Nonlinear Science*. Springer, New York (2003). Doi: https://doi.org/10.1007/978-0-387-21746-8
28. Xu, Q., Hesthaven, J.S.: Discontinuous Galerkin method for fractional convection–diffusion equations. SIAM J. Numer. Anal. **52**(1), 405–423 (2014). Doi: https://doi.org/10.1137/130918174
29. Yong, Z.H.O.U., Jinrong, W., Lu, Z.: *Basic Theory of Fractional Differential Equations*. World Scientific (2016)
30. Zhao, Z., Li, C.: Fractional difference/finite element approximations for the time–space fractional telegraph equation. Appl. Math. Comput. **219**(6), 2975–2988 (2012). Doi: https://doi.org/10.1016/j.amc.2012.09.022
31. Zheng, Y., Li, C., Zhao, Z.: A note on the finite element method for the space-fractional advection diffusion equation. Comput. Math. Appl. **59**(5), 1718–1726 (2010). Doi: https://doi.org/10.1016/j.camwa.2009.08.071
32. Zhuang, P., Liu, F., Anh, V., Turner, I.: New solution and analytical techniques of the implicit numerical method for the anomalous subdiffusion equation. SIAM J. Numer. Anal. **46**(2), 1079–1095 (2008). Doi: https://doi.org/10.1137/060673114
33. Zhuang, P., Liu, F., Anh, V., Turner, I.: Numerical methods for the variable-order fractional advection-diffusion equation with a nonlinear source term. SIAM J. Numer. Anal. **47**(3), 1760–1781 (2009). Doi: https://doi.org/10.1137/080730597

Chapter 5
Escaping from Current Minimizer by Using an Auxiliary Function Smoothed by Bezier Curves

Ahmet Sahiner, Idris A. Masoud Abdulhamid, and Nurullah Yilmaz

5.1 Introduction

The resolution of decision-making problems requires in the scientific field a quantification that directs the decision-making process to the optimal outcome. Numerous decisional problems that are encountered in the physical, chemical, economic, engineering, and other fields are often modeled as problems of optimization of a real function, which may, for example, represent the performance or cost of a system under certain conditions [1, 2]. This function is called as objective function which is generally non-linear and often subject to constraints which guarantee the acceptability of the solution being identified. The global optimization, in particular, has the following objectives:

- the analysis of non-linear decision models with multiple optimal solutions,
- the design and study of efficient resolution algorithms being able to identify the best overall solution.

From a mathematical perspective, the general problem of global optimization consists of identifying the global minimizer of a function $f : \Lambda \rightarrow R$ with $\Lambda \subset R^n$ [3]. The global optimization problem is indicated as follows:

$$f(x^*) = \min\{f(x) : x \in \Lambda\}. \tag{5.1}$$

A. Sahiner (✉) · N. Yilmaz
Department of Mathematics, Süleyman Demirel University, Isparta, Turkey
e-mail: ahmetsahiner@sdu.edu.tr; nurullahyilmaz@sdu.edu.tr

I. A. Masoud Abdulhamid
Süleyman Demirel University, Isparta, Turkey

© Springer Nature Switzerland AG 2020

J. A. T. Machado et al. (eds.), *Numerical Solutions of Realistic Nonlinear Phenomena*, Nonlinear Systems and Complexity 31, https://doi.org/10.1007/978-3-030-37141-8_5

Solving the global optimization problem means identifying the best of all solutions, and avoiding remaining trapped in a non-global local minimizer. The classes of problems to be solved in the context of optimization, both local and global, are varied and very different from each other [4]. These range from the problems of combinatorial optimization to quadratic programming, from concave minimization to convex differential programming and, from min-max problems to Lipschitz optimization. As a consequence, the proposed approaches for solving these problems are very varied. A very general strategy could be suitable in all cases, on the other hand, strictly specialized methods are applicable only for the class of problems for which they were designed.

Efficient algorithms that solve local optimization problems have been proposed for a long time ago: the steepest descent method, for example, dates back to Cauchy. However, these methods are not able to identify the global optimum of f, therefore, it becomes necessary to introduce appropriate global strategies that provide guarantees of convergence to the solution and perhaps exploit the efficiency of local strategies. The first algorithm for the resolution of the global optimization problem date back to the first half of the twentieth century [5, 6]. In the seventies, the first heuristic strategies were proposed, attempting, albeit with little success, to extend convergent algorithms in the one-dimensional case to higher dimensions. The first works on global optimization are attributable to Dixon and Szego [7]. Since then, there have been dozens of books and hundreds of articles on the subject published in numerous scientific journals. It should be noted that the proposed optimization methods, as numerical procedures, are generally not able to identify the exact solution of the problem, but only a numerical approximation of the global minimizer points, as well as the global minimizer itself. Depending on whether they use probabilistic elements, the methods developed for the resolution of the global optimization problem can be broadly divided into two large classes [8]:

- deterministic methods,
- stochastic methods.

Typically, all methods of the Branch and Bound [9], covering methods [10], space filling curves based methods [11, 12] belong to the class of deterministic methods, the same class also includes some methods that use perturbations of the objective function, such as the tunneling method and the filled function method (see, e.g., [13–41]). These methods provide an absolute guarantee of success but require restrictive assumptions about the function.

Stochastic methods generally require weaker hypotheses than the former and ensure probabilistic convergence to the global optimum. This class includes, for example, the two-phase methods, the random search methods, the simulated annealing methods, and the random direction methods [42–45]. There are in the literature methods based on stochastic procedures that, under particular conditions, are competitive from an empirical perspective but that do not provide guarantees of convergence. These methods, called heuristics, include all those approaches that simulate the processes of the biological evolution of a system, such as natural selection and apply them to populations of points which are recombined sequentially until the optimal solution is generated. Among these, we mention the evolutionary methods and genetic algorithms [46, 47]. The main difficulties in solving the global optimization problem are linked to the fact that the structure of the function to be minimized is often not known, so the number and distribution of the minimizer are not known. Moreover, as the size increases, the number of local minimizers also generally increases, with a consequent reduction in the attraction region of every minimizer and therefore the probability of identifying the global optimum decreases [48, 49]. The problems to be overcome in the application of stochastic methods based on local searches are as the following:

- avoid finding the same minimizers,
- have an efficient stopping criterion.

Our attention will focus exclusively on solving the global optimization problem in case the objective function f is continuous and the set Λ is a compact on $\mathbb{R}^n$.

The remainder of this article is prepared as follows: Sect. 5.2 introduces the important basic concepts and assumptions. Section 5.3 is devoted to describing the proposed new auxiliary function in detail with a literature review of the filled function method. The experiments on the performance of the offered algorithm are shown in Sect. 5.4. Finally, the conclusion of this article is outlined in Sect. 5.5.

5.2 Basic Concepts and Assumptions

Throughout the chapter, the symbol k represents the number of local minimizers, x_k^* represents the current local minimizer, and x^* represents the global minimizer of $f(x)$. Moreover, B_k^* represents the basin of $f(x)$ at the local minimizer x_k^*.

Let $f : \mathbb{R}^n \to \mathbb{R}$ be a real-valued function defined on some set $\Lambda \subset \mathbb{R}$. A point $x_k^* \in \Lambda$ is a local minimizer of f over Λ if there exists $\epsilon > 0$ such that $f(x) \geq f(x_k^*)$ for all $x \in \Lambda$ and $\|x - x_k^*\| < \epsilon$. A point $x^* \in \Lambda$ is a global minimizer of f if $f(x) \geq f(x^*)$ for all $x \in \Lambda$. If we replace "$\geq$" with "$>$" at above, then we define a strict local minimizer and a strict global minimizer, respectively.

Definition 5.1 ([17]) A basin of $f(x)$ at an isolated minimizer x_k^* is a connected domain $B(x_k^*)$ which contains x_k^* and in which starting from any point the steepest descent trajectory of $f(x)$ converges to x_k^*. but outside which the steepest descent trajectory of $f(x)$ does not converge to x_k^*. A hill of $f(x)$ at x_k^* is the basin of $-f(x)$ at its minimizer x_k^*, if x_k^* is a maximizer of $f(x)$.

Definition 5.2 ([17]) The direction $d \in \mathbb{R}^n$ is said to be a descent direction for $f : \mathbb{R}^n \to \mathbb{R}$ at $x \in \mathbb{R}^n$, if there exists $\epsilon > 0$ such that for all $\alpha \in (0, \epsilon]$

$$f(x + \alpha d_k) < f(x).$$

The definition of the filled function was first presented by Ge in [17]. The definition of the filled function is designed for various varieties. In this paper, we bring the following definition forward:

Definition 5.3 ([17]) A continuously differentiable function $H(x)$ is said to be a filled function of $f(x)$ at x_k^* , if it satisfies the following properties:

(i) x_k^* is a strict local maximum point of $H(x, x_k^*)$,
(ii) $H(x, x_k^*)$ has no stationary point in the set Λ_1,
(iii) If x_k^* is not a global minimizer of $f(x)$, then there exists a point x' such that it is a local minimizer of $H(x, x_k^*)$ on Λ_2.

In order to facilitate discussions, we assume that the following assumptions hold at the rest of this paper:

Assumption 1. The function f is Lipschitz continuous;
Assumption 2. The function $f : R^n \to R$ provides $f(x) \to +\infty$ as $\|x\| \to +\infty$; and
Assumption 3. The number of minimizers can be infinite, but the number of different function values at the minimizers is finite.

5.2.1 *Bezier Curves*

Bezier curve was developed by the French engineer Pierre Bezier (1910–1999) in 1962 for use in the design of Renault automobile bodies. Pierre Bezier then went on to develop the UNISURF CAD/CAM system. The Bezier curve is defined in terms of the locations of $n + 1$ points. These points are called data or control points. They form the vertices of what is called the control or Bezier characteristic polygon.

In general, a Bezier curve section can be fitted to any number of control points. The number of control points to be approximated determines the degree of the Bezier curve. For $n + 1$ control points, the Bezier curve is defined by the following polynomial of degree n:

$$P(u) = \sum_{j=0}^{n} Q_{j,n}(u) P_j,$$

where $0 \leq t \leq 1$ and $P(u)$ is any point on the curve and P_j is a control point, $Q_{j,n}(u)$ are the Bernstein polynomials. The Bernstein polynomial serves as the blending or basis function for the Bezier curve and is given by

$$Q_{j,n}(u) = C(n, j)u^j(1 - u)^{n-j},$$

where $C(n, j)$ is the binomial coefficient

$$C(n, j) = \frac{n!}{j!(n - j)!}.$$

A very useful property of a Bezier curve is that it always passes through the first and last control points. If we substitute $u = 0$ and 1, the boundary conditions at the two end points are

$$P(0) = P_0, \; P(1) = P_3.$$

The curve is tangent to the first and last segments of the characteristic polygon, the first derivatives when there are 4 control points ($n = 3$) are given by

$$P'(u) = -3(1-u)^2 P_0 + (3(1-u)^2 - 6u(1-u))P_1 + (6u(1-u) - 3u^2)P_2 + 3u^2 P_3.$$

Therefore the tangent vectors at the starting and ending points are

$$P'(0) = 3(P_1 - P_0),$$

$$P'(1) = 3(P_3 - P_2).$$

Similarly, it can be shown that the second derivative at P_0 is determined by P_0, P_1 and P_2 or in general, the r-th derivative at an endpoint is determined by its r neighboring vertices.

5.3 The Auxiliary Function Method

5.3.1 Literature Review of the Filled Function

In 1983, Ren Pu Ge conferred the concept of the filled function method [4]. His idea was to construct an auxiliary function at the current minimizer point x_k^* so that it would become a strict maximizer of that auxiliary function and its basin B_k^* be a part of a hill of the auxiliary function [5].

Ge particularly purposed the filled function of f at a local minimizer x_k^* on Λ as follows:

$$FF(x, x_k^*, \beta, \tau) = \frac{1}{\beta + f(x)} \exp\left(-\frac{\|x - x_k^*\|^2}{\tau^2}\right), \tag{5.2}$$

where β and τ must be chosen properly. In 1987, Ge [17] developed the filled function (5.2) such that rewritten as:

$$\tilde{F}F(x, x_k^*, \beta, \tau) = \frac{1}{\beta + f(x)} \exp\left(-\frac{\|x - x_k^*\|}{\tau^2}\right). \tag{5.3}$$

In the same period of time, Qin and Ge [17] introduced a series of filled functions that led to a great leap in the optimization field:

$$FF_1(x, x_k^*, \beta, \tau) = -\tau^2 \ln[\beta + f(x)] - \|x - x_k^*\|^2, \tag{5.4}$$

$$FF_2(x, x_k^*, \beta, \tau) = -\tau^2 \ln[\beta + f(x)] - \|x - x_k^*\|, \tag{5.5}$$

$$Q_1(x, x_k^*, a) = [f(x_k^*) - f(x)] \exp(a\|x - x_k^*\|^2), \tag{5.6}$$

$$Q_2(x, x_k^*, a) = [f(x_k^*) - f(x)] \exp(a\|x - x_k^*\|), \tag{5.7}$$

$$E_1(x, x_k^*, a) = 2a[f(x_k^*) - f(x)](x - x_k^*) - \nabla f(x), \tag{5.8}$$

$$E_2(x, x_k^*, a) = a[f(x_k^*) - f(x)]\frac{(x - x_k^*)}{\|x - x_k^*\|} - \nabla f(x). \tag{5.9}$$

In 1999, Wu [41] postulated that the filled functions (5.8) and (5.9) failed to exist for the general objective function $f(x)$ and for $x \in \mathbb{R}^2$. Moreover, it is obvious that (5.6) and (5.7), nevertheless, result from the related numerical problem, Wu proposed the following filled functions:

$$U(x, x_k^*, \rho) = [f(x_k^*) - f(x)](\|x - x_k^*\|^2 + 1)^\rho, \tag{5.10}$$

$$U(x, x_k^*, \rho) = -[f(x) - f(x_k^*)](\|x - x_k^*\| + 1)^\rho, \tag{5.11}$$

where $\rho > 0$ is a sufficiently large natural number. In 1992, Sheng and Wang [24] proposed the following filled function:

$$K(x, x_k^*, \beta, \tau) = \ln\left(\frac{1}{\beta + f(x)} + 1\right) \exp\left(-\frac{\|x - x_k^*\|}{\tau^2}\right). \tag{5.12}$$

This was another variant of (5.2). In fact, the majority of filled functions proposed in 2000 were simply varied or generalizations of (5.2) and (5.6). In 1994, Zuang [6] generalized (5.2) and proposed the following filled function:

$$D_1(x, x_1^*, \beta, \tau) = H(\beta + f(x)) \exp\left(-\frac{\|x - x_k^*\|}{\tau^2}\right), \tag{5.13}$$

where H is a twice continuously differentiable, strictly decreasing function which satisfies $[H'(\lambda)]^2 \leq H(\lambda)H'')(\lambda)$, for all $\lambda > 0$. In 2001, Xu et al. [20] developed the filled function at (5.11) and produced the next two filled functions:

$$D_2(x, x_k^*, \beta, a) = \zeta(\beta + f(x)) \exp[a\eta(\|x - x_k^*\|^\alpha)], \tag{5.14}$$

$$D_3(x, x_k^*, a) = -\psi([f(x) - f(x_k^*)]) \exp(a\eta(\|x - x_k^*\|^\alpha)), \tag{5.15}$$

where $\alpha > 0$ and a satisfy some conditions and the functions ζ, η and ψ as well as the parameter β.

Lately, Xu et al. [20] qualified their filled function (5.15) and suggested the following filled function:

$$D_4(x, x_k^*, a) = -\phi(f(x) - f(x_k^*)) - a\varphi(\|x - x_k^*\|^{\alpha}), \quad \alpha(> 0) \in \mathbb{Z}, \tag{5.16}$$

where the function ϕ and φ satisfy some conditions. Another modification of (5.6) was inferred via Kong [19] in 2000, thus

$$P(x, x_k^*, a) = -\ln[f(x) - f(x_k^*) + 1] \exp(a\|x - x_k^*\|^2), \tag{5.17}$$

where $a > 0$ is sufficiently large.

Recently, several researchers have persisted in introducing a number of auxiliary functions, some of which can be classified as filled functions. In 2012, Sahiner A., and Gökkaya, H. constructed a filled function for non-smooth global optimization [21]. In 2017, Sahiner et al. introduced a new methodology for modeling the given data and finding the global optimum value of the model function [22]. In 2018, Sahiner, A., and Shehab, S. introduced a new global optimization technique based on the directional search for unconstrained optimization problem [23].

5.3.2 A New Auxiliary Function and Its Properties

In order to find a global minimizer of the objective function f, the main subject of the auxiliary function method is to obtain the lowest minimizer of f or verify whether the concerned local minimizer is a global minimizer of f. This gradually relies on the execution of the auxiliary function.

In this section, we introduce a new auxiliary function for the problem (5.1) at the current local minimizer x_k^*, for that, we first suggest the function:

$$H(x, x_k^*) = \min\{f(x), f(x_k^*)\}.$$

The function $H(x, x_k^*)$ is eliminating the local minimizers which are worse than the best one obtained so far. By using this function, we can decrease the number of local minimizers, which will be extremely essential to the optimization algorithms. It is evident that for $\forall x \in \Lambda$, the function $H(x, x_k^*)$ has the two properties:

1. $H(x, x_k^*) = f(x_k^*)$ if $f(x) \geq f(x_k^*)$,
2. $H(x, x_k^*) = f(x)$ if $f(x) < f(x_k^*)$.

The function $H(x, x_k^*)$ can be rewritten by using a multiplication with a piecewise function $\mathscr{D}_{\mathscr{U}_k}(t)$ as the following:

$$H(x, x_k^*) = f(x_k^*) + (f(x) - f(x_k^*))\mathscr{D}_{\mathscr{U}_k}(t), \tag{5.18}$$

where $\mathscr{U}_k = \{x \in \mathbb{R}^n : f(x) < f(x_k^*)\}$ and $\mathscr{D}_{\mathscr{U}_k}(t) : \mathbb{R}^n \to \mathbb{R}$ can be defined as:

$$\mathscr{D}_{\mathscr{U}_k}(t) = \begin{cases} 1, & t \geq 0 \\ |t^{3/2}|, & t < 0. \end{cases} \tag{5.19}$$

To make the function $H(x, x_k^*)$ smooth, it is sufficient to make $\mathscr{D}_{\mathscr{U}_k}(t)$ smooth, for this, the continuously differentiable functions $\eta_1(t, \beta)$ and $\eta_2(t, \beta)$ were created by omitting the parameter u from the Bernstein basis polynomials as follows:

$$\eta_1(t, \beta) = \left(\frac{\frac{2\left(3\beta - \frac{6}{5}\right)^2}{9\left(\beta - \frac{3}{10}\right)^2} - 2}{y_1} + 2y_1 + \frac{2\beta - \frac{4}{5}}{\beta - \frac{3}{10}} + 1 \right) \left(\frac{\beta - \frac{2}{5}}{\beta - \frac{3}{10}} + \frac{y_3}{y_1} + y_1 - 1 \right)^2,$$

and

$$\eta_2(t, \beta) = \left(y_4 - \frac{3}{4y_4} + \frac{1}{2} \right)^2 \left(\frac{3}{2y_4} - 2y_4 + 2 \right),$$

where

$$y_1 = \left(\frac{(3\beta - \frac{6}{5})^3}{27(\beta - \frac{3}{10})^3} - \frac{t}{2\beta - \frac{3}{5}} + \sqrt{\left(\frac{t}{2\beta - \frac{3}{5}} - \frac{(3\beta - \frac{6}{5})^3}{27\left(\beta - \frac{3}{10}\right)^3} + y_2 \right)^2 - y_3^3} - y_2 \right)^{\frac{1}{3}},$$

$$y_2 = \frac{\left(3\beta - \frac{6}{5}\right)\left(3\beta - \frac{9}{10}\right)}{6\left(\beta - \frac{3}{10}\right)^2},$$

$$y_3 = \frac{\left(3\beta - \frac{6}{5}\right)^2}{9\left(\beta - \frac{3}{10}\right)^2} - 1,$$

$$y_4 = \left(\sqrt{\left(\frac{5\beta}{2} + \frac{5t}{2} + \frac{5}{8} \right)^2 + \frac{27}{64}} - \frac{5t}{2} - \frac{5\beta}{2} - \frac{5}{8} \right)^{\frac{1}{3}}.$$

Figure 5.1 shows the behavior of the functions $\eta_1(t, \beta)$ and $\eta_2(t, \beta)$.

The function $H(x, x_k^*)$ after smoothed $\mathscr{D}_{\mathscr{U}_k}(t)$ becomes

$$\tilde{H}(x, x_k^*) = f(x_k^*) + (f(x) - f(x_k^*))\tilde{\mathscr{D}}_{\mathscr{U}_k}(t, \beta), \tag{5.20}$$

where $\tilde{\mathscr{D}}_{\mathscr{U}_k}(t)$ is a smooth function that can be defined as the following form based on η_1 and η_2:

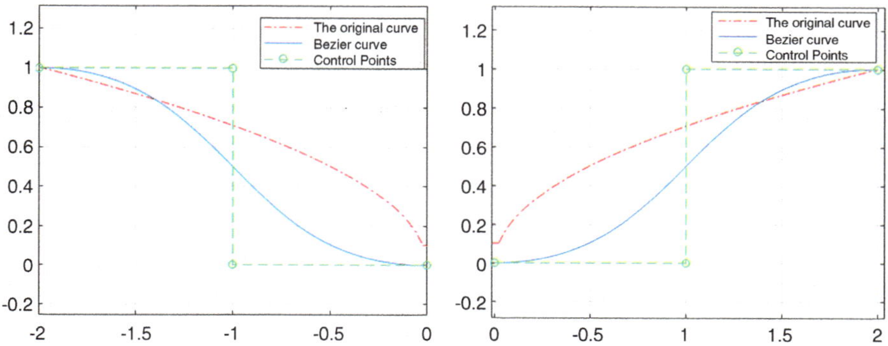

Fig. 5.1 Curves with blue color represent the behavior of the functions $\eta_1(t, \beta)$ and $\eta_2(t, \beta)$. Together they offer the smooth version of $\mathscr{D}_{\mathscr{U}_k}(t)$ (with the red color) after being smoothed using Bezier curves

$$\tilde{\mathscr{D}}_{\mathscr{U}_k}(t, \beta) = \begin{cases} 1, & t > 0, \\ \eta_1(t, \beta), & 0 \geq t > -\beta, \\ \eta_2(t, \beta), & -\beta \geq t > -0.5 - \beta, \\ 1, & t \leq -0.5 - \beta, \end{cases} \tag{5.21}$$

where $t = f(x) - f(x_k^*)$ and $\beta > 0$.

Lemma 5.1 *Suppose that x_k^* is a local minimizer of $f(x)$, we have*

$$\lim_{\beta \to 0} \tilde{H}(x, x_k^*, \beta) = H(x, x_k^*).$$

During the elimination process by the function $H(x, x_k^*)$, there will be a lot of lost information that because of the flatland as a result of eliminating the local minimizer higher than the best one obtained so far. This causes the inability to move from the current local minimizer to a better one which requires appending an additional function $\psi(\rho)$ to the auxiliary function $\tilde{H}(x, x_k^*, \beta)$. Examples of the function $\psi(\rho)$ in the literature, $\psi(\rho) = \frac{1}{\rho^m}$, $(m > 1)$; $\psi(\rho) = \frac{1}{\ln(1+\rho)}$; $\psi(\rho) = \frac{1}{\arctan(\rho)}$; and $\psi(\rho) = \exp(\rho)$. Therefore, we added $\exp\left(\frac{-\|x - x_k^*\|^2}{\tau}\right)$ as a term of $\tilde{H}(x, x_k^*, \beta)$. Thus, $\tilde{H}(x, x_k^*, \beta)$ in its final form become as follows:

$$\tilde{H}(x, x_k^*, \beta, \tau) = f(x_k^*) + (f(x) - f(x_k^*))\tilde{\mathscr{D}}_{\mathscr{U}_k}(t, \beta) + \exp\left(\frac{-\|x - x_k^*\|^2}{\tau}\right). \tag{5.22}$$

Based on what has been explained above, the auxiliary function $\tilde{H}(x, x_k^*, \beta, \tau)$ can be redefined as follows:

Definition 5.4 $\tilde{H}(x, x_k^*, \beta, \tau)$ is called a filled function if it achieves

1. $\tilde{H}(x, x_k^*, \beta, \tau)$ has at least one local minimizer x_k^*,
2. $\tilde{H}(x, x_k^*, \beta, \tau)$ has no stationary point in the set

$$\Lambda_1 = \{x : f(x_k^*) \leq f(x), x \in \Lambda/\{x_k^*\}\},$$

3. $\tilde{H}(x, x_k^*, \beta, \tau)$ has a minimizer in the set:

$$\Lambda_2 = \{x : f(x_k^*) > f(x), x \in \Lambda\},$$

if x_k^* is not a global minimizer of $f(x)$.

Theorem 5.1 *Suppose that x_k^* is the current local minimizer of $f(x)$, then x_k^* is the local maximizer of $\tilde{H}(x, x_k^*, \beta, \tau)$.*

Proof Since x_k^* is the current local minimizer of $f(x)$, then there exists a neighborhood $\xi = Q(x_k^*, \sigma)$ of x_k^* and $\sigma > 0$ such that $f(x) \geq f(x_k^*)$ for all $x \in \xi$ and $x \neq x_k^*$, then

$$\frac{\tilde{H}(x, x_k^*, \beta, \tau)}{\tilde{H}(x_k^*, x_k^*, \beta, \tau)} = \frac{f(x_k^*) + \exp\left(\frac{-\|x - x_k^*\|^2}{\tau}\right)}{f(x_k^*) + 1} < 1.$$

Therefore, we have

$$\tilde{H}(x, x_k^*, \beta, \tau) < \tilde{H}(x_k^*, x_k^*, \beta, \tau).$$

Thus, x_k^* is a local maximizer of $\tilde{H}(x, x_k^*, \beta, \tau)$.

Theorem 5.2 *Suppose that f is continuously differentiable function. If x_k^* is a local minimizer of f, then $\tilde{H}(x, x_k^*, \beta, \tau)$ has no stationary point for $x \in \Lambda_1$.*

Proof In the case of $t = f(x) - f(x_k^*) > \beta$, we have

$$\tilde{H}(x, x_k^*, \beta, \tau) = f(x_k^*) + \exp\left(\frac{-\|x - x_k^*\|^2}{\tau}\right).$$

For any x satisfying $f(x) \geq f(x_k^*)$, and $x \neq x_k^*$ we have

$$\nabla \tilde{H}\gamma(x, x_k^*, \beta, \tau) = -2\frac{(x - x_k^*)}{\tau} \exp\left(\frac{-\|x - x_k^*\|^2}{\tau}\right).$$

Therefore, $\nabla \tilde{H}(x, x_k^*, \beta, \tau) \neq 0$. Consequently, $\tilde{H}(x, x_k^*, \beta, \tau)$ cannot have a stationary point for $x \in \Lambda_1$.

Theorem 5.3 *Suppose that* x_k^* *is a local minimizer of* $f(x)$*, but not a global minimizer. Then there exists a point* $\hat{x} \in \Lambda_2$*, such that* $\hat{x}$ *is a local minimizer of* $\tilde{H}(x, x_k^*, \beta, \tau)$.

Proof First of all, we will prove that the function $\tilde{H}(x, x_k^*, \beta, \tau)$ contains a local minimizer point. For this, assume that f has a local minimizer point x_{k+1}^* lower than x_k^*. Then, we achieve that the subset $\Gamma = \{x : f(x_k^*) \geq f(x), x \in \Lambda\}$ is not empty, and because of the continuity and boundedness of f, the subset Γ is closed. Then $\tilde{H}(x, x_k^*, \beta, \tau)$ has a local minimizer on Γ.

Now, suppose that x_{k+1}^* is a local minimizer point of f lower than x_k^*, and $\hat{x}$ is the local minimizer of $\tilde{H}(x, x_k^*, \beta, \tau)$, then the gradient of $\tilde{H}(x, x_1^*, \beta, \tau)$ is equal to zero. Therefore,

$$\exp\left(\frac{-\|x - x_k^*\|^2}{\tau}\right) + f(\hat{x}) = 0$$

accordingly, $-\exp\left(\frac{-\|x-x_k^*\|^2}{\tau^2}\right) = f(\hat{x})$. Thus, we can write

$$(\hat{x} - x_k^*)\nabla f(\hat{x}) = (\hat{x} - x_k^*)\left(-\exp\left(\frac{-\|x - x_k^*\|^2}{\tau}\right)\right) > 0.$$

Since both vectors $(x_{k+1}^* - x_k^*)$ and $(\hat{x} - x_k^*)$ are almost equal for a small value for the real parameter a and $\|x_{k+1}^* - x_k^*\| \leq \|\hat{x} - x_k^*\|$, we can understand that they are in the same direction. This confirms that $(\hat{x} - x_k^*)(-\exp(\frac{-\|x-x_k^*\|^2}{\tau})) > 0$.

Algorithm

Step 0. Let $k = 0$, $\tau = 0.5$, $\beta = 0.5$, $\epsilon = 0.001$ and select an initial point $x_0 \in \Lambda$.
Step 1. Let M be the number of directions d_k for $k = 1, 2, 3, \ldots.., M$.
Step 2. Construct the one-dimensional function $Y(\theta) = f(x_0 + \theta d_k)$.
Step 3. 1. From any starting point θ_0, find a local minimizer θ_k^j for $Y(\theta)$ and select $\lambda = -1$.
2. Construct the auxiliary function $\tilde{H}_\theta(\theta, \theta_k^j, \beta, \tau)$ at θ_k^j.
3. Beginning from $\theta_0 = \theta_k^j + \lambda\epsilon$, find the minimizer θ_H of $\tilde{H}_\theta(\theta, \theta_k^j, \beta, \tau)$.
4. If $\theta_H \in \Lambda$, go to (5.5); otherwise go to (5.7).
5. Beginning from θ_H, minimize $Y(\theta)$ to find θ_k^{j+1} lower than θ_k^j and go to (5.6).
6. If $\theta_k^{j+1} \in \Lambda$, put $\theta_k^j = \theta_k^{j+1}$ and go to (5.2).
7. If $\lambda = 1$, stop and put $\theta_k^* = \theta_k^j$; otherwise, take $\lambda = 1$ and go to (5.3).

Step 4. Using $\hat{x}_k = x_0 + \theta_k^* d_k$, calculate $\hat{x}_k$.

Step 5. Using $\hat{x}_k$ as the initial point, find x^* of the function $f(x)$.
Step 6. If $k < M$, let $k = k + 1$; create a new search direction d_{k+1}, and go back to Step 2; otherwise, go to Step 7.
Step 7. Select the global minimizer of $f(x)$ which is determined by

$$f(x^*) = \min\{f(x_k^*)\}_{k=1}^{M}.$$

5.4 Numerical Experiments

In this section, we present some examples to test the numerical experiments to illustrate the efficiency of the algorithm. MATLAB was used to code the algorithm, and the directional method was used to find the local minimum points for the test problems. From the experimental results, it appears that the algorithm works well and did not prove any important performance variations. The stopping criterion for the algorithm was $\|\nabla f\| \leq 10^{-4}$. The test functions for the numerical experiments were obtained from [5, 8, 22–24] as presented below:

Problem 1 (Two-Dimensional Function) This two-dimensional function appears in article [5]. It is defined as

$$\min_{x\in\mathbb{R}^n} f(x) = (1 - 2x_2 + c\sin(4\pi x_2) - x_1)^2 + (x_2 + 0.5\sin(2\pi x_1))^2,$$

the problem has several local minimizer and many global minima with $f(x^*) = 0$ in the domain $[0, 10] \times [-10, 0]$. We began from the initial point $x_0 = (1.5, 1.5)$ for $c = 0.2$, $x_0 = (2.4, 2.4)$ for $c = 0.05$, and from $x_0 = (-1.3, -1.3)$ for $c = 0.5$ and we used a directional search method to minimize the objective function f. The problem has three local minimizer, one of which is the global minimizer at $x^* = (0, 0)$ with $f(x^*) = 0$. The proposed filled function method achieved recognizing the global minimizer. The time to reach the global minimizer was 0.763825 seconds. The problem was tested for c = 0.2, 0.05, and 0.5, respectively. The detailed results are listed in Tables 5.1, 5.2, and 5.3.

Problem 2 (Three-Hump Camel Function)

$$\min_{x\in\mathbb{R}^n} f(x) = 2x_1^2 - 1.05x_1^4 + \frac{x_1^6}{6} + x_1x_2 + x_2^2,$$

the global minimizer is $x^* = (0, 0)$ with $f(x^*) = 0$. Table 5.4 lists the numerical results obtained for Problem 2.

Table 5.1 Results obtained for Problem 1 with $c = 0.2$

k	x_k	$f(x_k)$	x_1^*	$f(x_1^*)$
1	$\begin{pmatrix}1.5000\\1.5000\end{pmatrix}$	14.5000	$\begin{pmatrix}-1.6756\\0.9180\end{pmatrix}$	0.6688
2	$\begin{pmatrix}1.1000\\1.1000\end{pmatrix}$	5.1010	$\begin{pmatrix}1.0175\\0.0548\end{pmatrix}$	2.4568×10^{-14}
3	$\begin{pmatrix}0.7000\\0.7000\end{pmatrix}$	2.3471	$\begin{pmatrix}0.4091\\0.2703\end{pmatrix}$	6.7978×10^{-14}
4	$\begin{pmatrix}0.3000\\0.3000\end{pmatrix}$	0.0311	$\begin{pmatrix}0.4091\\0.2703\end{pmatrix}$	1.3487×10^{-15}

Table 5.2 Results obtained for Problem 1 with $c = 0.05$

k	x_k	$f(x_k)$	x_1^*	$f(x_1^*)$
1	$\begin{pmatrix}-1.3000\\-1.3000\end{pmatrix}$	24.9786	$\begin{pmatrix}-0.7300\\0.7934\end{pmatrix}$	0.1022
2	$\begin{pmatrix}-1.4000\\-1.4000\end{pmatrix}$	28.7603	$\begin{pmatrix}0.1487\\0.4021\end{pmatrix}$	1.332×10^{-13}

Table 5.3 Results obtained for Problem 1 with $c = 0.5$

k	x_k	$f(x_k)$	x_1^*	$f(x_1^*)$
1	$\begin{pmatrix}2.4000\\2.4000\end{pmatrix}$	48.9984	$\begin{pmatrix}-2.7229\\1.3765\end{pmatrix}$	1.0019
2	$\begin{pmatrix}1.9000\\1.9000\end{pmatrix}$	31.5993	$\begin{pmatrix}1.0568\\0.1746\end{pmatrix}$	2.6366×10^{-13}
3	$\begin{pmatrix}1.4000\\1.4000\end{pmatrix}$	14.7330	$\begin{pmatrix}-2.7229\\1.3765\end{pmatrix}$	1.0019
4	$\begin{pmatrix}0.9000\\0.9000\end{pmatrix}$	6.1583	$\begin{pmatrix}1.0000\\0.0000\end{pmatrix}$	1.9637×10^{-14}
5	$\begin{pmatrix}0.4000\\0.4000\end{pmatrix}$	0.4676	$\begin{pmatrix}0.1026\\0.3005\end{pmatrix}$	2.7425×10^{-15}

Problem 3 (Six-Hump Back Camel Function)

$$\min_{x\in\mathbb{R}^n} f(x) = 4x_1^2 - 2.1x_1^4 + \frac{1}{3}x_1^6 - x_1x_2 - 4x_2^2 + 4x_2^4,$$

the global minimizer is one of the points $(0.089842, 0.71266)$ or $(-0.089842, -0.71266)$ with $f(x^*) = -1.031628$ in the domain $[-3, 3] \times [-3, 3]$. Table 5.5 lists the numerical results obtained for Problem 3.

Table 5.4 Results obtained for Problem 2

k	x_k	$f(x_k)$	x_1^*	$f(x_1^*)$
1	$\begin{pmatrix}-0.3000\\-0.3000\end{pmatrix}$	0.1716	$\begin{pmatrix}-0.1190\\-0.5363\end{pmatrix} \times 10^{-06}$	2.5215×10^{-13}
2	$\begin{pmatrix}-0.5000\\-0.5000\end{pmatrix}$	0.4370	$\begin{pmatrix}-0.0279\\-0.5301\end{pmatrix} \times 10^{-07}$	2.6774×10^{-15}
3	$\begin{pmatrix}-0.7000\\-0.7000\end{pmatrix}$	0.7475	$\begin{pmatrix}0.2619\\-0.0793\end{pmatrix} \times 10^{-07}$	1.2266×10^{-15}
4	$\begin{pmatrix}-0.9000\\-0.9000\end{pmatrix}$	1.0197	$\begin{pmatrix}-0.0139\\-0.2612\end{pmatrix} \times 10^{-06}$	6.4964×10^{-14}

Table 5.5 Results obtained for Problem 3

k	x_k	$f(x_k)$	x_1^*	$f(x_1^*)$
1	$\begin{pmatrix}-0.5000\\-0.5000\end{pmatrix}$	-0.1260	$\begin{pmatrix}-0.0898\\-0.7127\end{pmatrix}$	−1.0316

Table 5.6 Results obtained for Problem 4

k	x_k	$f(x_k)$	x_1^*	$f(x_1^*)$
1	$\begin{pmatrix}-0.9000\\-0.9000\end{pmatrix}$	1.2970	$\begin{pmatrix}-0.0579\\-0.2486\end{pmatrix} \times 10^{-07}$	7.5233×10^{-16}
2	$\begin{pmatrix}-1.1000\\-1.1000\end{pmatrix}$	2.1901	$\begin{pmatrix}-2.0000\\0.0000\end{pmatrix}$	9.6088×10^{-15}

Problem 4 (Treccani Function)

$$\min_{x \in \mathbb{R}^n} f(x) = x_1^4 + 4x_1^3 + 4x_1^2 + x_2^2,$$

it is clear that the global minimizers are $(-2, 0)$ and $(0, 0)$ with $f(x^*) = 0$ in the domain $[-3, 3] \times [-3, 3]$. Table 5.6 lists the numerical results obtained for Problem 4.

Problem 5 (Goldstein and Price Function)

$$\min_{x \in \mathbb{R}^n} f(x) = g(x)h(x),$$

where

$$g(x) = 1 + (x_1 + x_2 + 1)^2(19 - 14x_1 + 3x_1^2 - 14x_2 + 6x_1x_2 + 3x_2^2),$$

Table 5.7 Results obtained for Problem 5

k	x_k	$f(x_k)$	x_1^*	$f(x_1^*)$
1	$\begin{pmatrix} -1.5000 \\ -1.5000 \end{pmatrix}$	1.1186×10^{04}	$\begin{pmatrix} 0.0000 \\ -1.0000 \end{pmatrix}$	3.0000

Table 5.8 Results obtained for Problem 6

k	x_k	$f(x_k)$	x_1^*	$f(x_1^*)$
1	$\begin{pmatrix} -4.3000 \\ -4.3000 \end{pmatrix}$	716.4200	$\begin{pmatrix} 1.0000 \\ 3.0000 \end{pmatrix}$	1.4771×10^{-13}

Table 5.9 Results obtained for Problem 7

k	x_k	$f(x_k)$	x_1^*	$f(x_1^*)$
1	$\begin{pmatrix} -34.2000 \\ -34.2000 \\ -34.2000 \\ -34.2000 \\ -34.2000 \\ -34.2000 \end{pmatrix}$	1.5860×10^{03}	$\begin{pmatrix} 6.0000 \\ 10.0000 \\ 12.0000 \\ 12.0000 \\ 10.0000 \\ 6.0000 \end{pmatrix}$	−50.000

and

$$h(x) = 30 + (2x_1 - 3x_2)^2(18 - 32x_1 + 12x_1^2 - 48x_2 - 36x_1x_2 + 27x_2^2).$$

The global minimizer is $(0, -1)$ with $f(x^*) = 3$ in the domain $[-3, 3] \times [-3, 3]$. Table 5.7 lists the numerical results obtained for Problem 5.

Problem 6 (Booth Function)

$$\min_{x \in \mathbb{R}^n} f(x) = (x_1 + 2x_2 - 7)^2 + (2x_1 + x_2 - 5)^2,$$

the global minimizer is $(1, 3)$ with $f(x^*) = 0$ in the domain $[-10, 10] \times [-10, 10]$. Table 5.8 lists the numerical results obtained for Problem 6.

Problem 7 (Trid Function)

$$\min_{x \in \mathbb{R}^n} f(x) = \sum_{i=1}^{n} (x_i - 1)^2 - \sum_{i=2}^{n} x_i x_{i-1},$$

the global minimizer is at $f(x^*) = -50$ in the domain $[-n^2, n^2]^n$. Table 5.9 lists the numerical results obtained for Problem 7.

Table 5.10 Results obtained for Problem 8

k	x_k	$f(x_k)$	x_1^*	$f(x_1^*)$
1	$\begin{pmatrix} -2.2000 \\ -2.2000 \\ -2.2000 \\ -2.2000 \\ -2.2000 \\ -2.2000 \\ -2.2000 \\ -2.2000 \\ -2.2000 \\ -2.2000 \end{pmatrix}$	1.2181	$\begin{pmatrix} -0.0316 \\ 0.0032 \\ -0.0127 \\ -0.0127 \\ -0.0316 \\ 0.0032 \\ -0.0127 \\ -0.0127 \\ -2.2000 \\ -2.2000 \end{pmatrix}$	3.8940×10^{-6}
2	$\begin{pmatrix} -2.4000 \\ -2.4000 \\ -2.4000 \\ -2.4000 \\ -2.4000 \\ -2.4000 \\ -2.4000 \\ -2.4000 \\ -2.4000 \\ -2.4000 \end{pmatrix}$	1.4603	$\begin{pmatrix} -0.0242 \\ 0.0024 \\ -0.0075 \\ -0.0075 \\ -0.0242 \\ 0.0024 \\ -0.0075 \\ -0.0075 \\ -2.4000 \\ -2.4000 \end{pmatrix}$	1.7654×10^{-06}

Problem 8 (Powell Function)

$$\min_{x\in\mathbb{R}^n} \sum_{i=1}^{n/4} \left[(x_{4i-3} + 10x_{4i-2})^2 + 5(x_{4i-1} - x_{4i})^2 + (x_{4i-2} - 2x_{4i-1})^4 \right.$$
$$\left. +10(x_{4i-3} - x_{4i})^2\right],$$

the global minimizer is $(0, \ldots, 0)$ with $f(x^*) = 0$ in the domain $[-4, 5]^n$. Table 5.10 lists the numerical results obtained for Problem 8.

Problem 9 (Rastrigin Function)

$$\min_{x\in\mathbb{R}^n} f(x) = 10n + \sum_{i=1}^{n}(x_i^2 - 10\cos(2\pi x_i)),$$

the global minimizer is $(0, .., 0)$ with $f(x^*) = 0$ in the domain $[-5.12, 5.12]$. Table 5.11 lists the numerical results obtained for Problem 9.

Table 5.11 Results obtained for Problem 9

k	x_k	$f(x_k)$	x_1^*	$f(x_1^*)$
1	$\begin{pmatrix} -0.5200 \\ -0.5200 \\ -0.5200 \\ -0.5200 \\ -0.5200 \\ -0.5200 \\ -0.5200 \\ -0.5200 \\ -0.5200 \\ -0.5200 \end{pmatrix}$	201.9155	$\begin{pmatrix} -0.9950 \\ -0.9950 \\ -0.9950 \\ -0.9950 \\ -0.9950 \\ -0.9950 \\ -0.9950 \\ -0.9950 \\ -0.9950 \\ -0.9950 \end{pmatrix}$	9.9496
2	$\begin{pmatrix} -0.0200 \\ -0.0200 \\ -0.0200 \\ -0.0200 \\ -0.0200 \\ -0.0200 \\ -0.0200 \\ -0.0200 \\ -0.0200 \\ -0.0200 \end{pmatrix}$	0.7925	$\begin{pmatrix} -0.6942 \\ -0.6942 \\ -0.6906 \\ -0.6942 \\ -0.6942 \\ -0.6942 \\ -0.6942 \\ -0.6942 \\ -0.6942 \\ -0.6942 \end{pmatrix} \times 10^{-8}$	1.2790×10^{-13}

5.5 Conclusion

In this paper, we introduced a new algorithm which has the ability to convert the objective function into a one-dimensional function using an auxiliary function smoothed with the assistance of Bezier curves, followed by finding the global minimizer of the given objective function. MATLAB program has been used to obtain the numerical results to verify the effectiveness and efficiency of the proposed method. Since the functions obtained by using the Bernstein polynomial are often extremely long and complicated especially in the case of higher numbers of control points, we applied the MUPAD program to improve and simplify those equations.

References

1. Resener, M., Haffner, S., Pereira, L.A., Pardalos, P.A.: Optimization techniques applied to planning of electric power distribution systems: a bibliographic survey. Energy Syst. **9**, 473–509 (2018)
2. Addis, B., Cassioli, A., Locatelli, M., Schoen, F.: A global optimization method for the design of space trajectories. Comput. Optim. Appl. **48**, 635–652 (2011)
3. Floudas, C.: *Encyclopedia of Optimization*. Springer, New York (2009)

4. Ge, R.P: A filled function method for finding a global minimizer of a function of several variables. Math. Program. **46**, 191–204 (1990)
5. Ng, C.K.: High performance continuous/discrete global optimization methods. Doctoral dissertation, Chinese University of Hong Kong (2003)
6. Zhuang, J.N.: A generalized filled function method for finding the global minimizer of a function of several variables. Num. Math. J. Chinese Univ. **16**(3), 279–287 (1994)
7. Dixon, L., Szego, G.: *The Global Optimization Problem. An Introduction. Toward Global Optimization*. North-Holland, Amsterdam (1978)
8. Shang, Y., He, X., Lei, H.: Electromagnetism-like mechanism algorithm for global optimization. J. Comput. Appl. **30**, 2914–2918 (2010)
9. Land, A.H., Doig, A.G.: An automatic method of solving discrete programming problems. Econometrica **28**, 497–520 (1960)
10. Jones, D.R., Perttunen, C.D., Stuckman, B.E.: Lipschitzian optimization without the Lipschitz constant. J. Optimiz. Theory Appl. **79**, 157–181 (1993)
11. Lera, D., Sergeyev, Y.D.: Lipschitz and Hölder global optimization using space-filling curves. Appl. Numer. Math. **60**, 115–129 (2010)
12. Lera, D., Sergeyev, Y.D.: Deterministic global optimization using space-filling curves and multiple estimates of Lipschitz and Hölder constants. Commun. Nonlinear Sci. Numer. Simulat. **23**, 328–342 (2015)
13. Ge, R.P.: The theory of filled function method for finding global minimizers of nonlinearly constrained minimization problem. J. Comput. Math. **5**, 1–9 (1987)
14. Zhang, L., Ng, C., Li, D., Tian, W.: A new filled function method for global optimization. J. Global Optim. **28**, 17–43 (2004)
15. Wang, C., Yang, Y., Li, J.: A new filled function method for unconstrained global optimization. J. Comput. Appl. Math. **225**, 68–79 (2009)
16. Ng, C.K., Zhang, L.S., Li, D., Tian, W.W.: Discrete filled function method for discrete global optimization. Comput. Optim. Appl. **31**(1), 87–115 (2005)
17. Ge, R., Qin, Y.: A class of filled functions for finding global minimizers of a function of several variables. J. Optimiz. Theory App. **54**, 241–252 (1987)
18. Wu, Z., Lee, H., Zhang, L., Yang, X.: A novel filled function method and quasi-filled function method for global optimization. Comput. Optim. Appl. **34**, 249–272 (2005)
19. Min, K.: On the filled function method for nonsmooth program. J. Syst. Sci. Math. Sci. **20**(4), 148–154 (2000)Ŕ
20. Zheng, X., Chengxian, X.: Filled functions for unconstrained global optimization. Appl. Math. Ser. B. **15**, 307–318 (2000)
21. Sahiner, A., Gokkaya, H., Yigit, T.: A new filled function for nonsmooth global optimization. In: AIP Conference Proceedings, Vol. 1479(1), pp. 972–974 (2012)
22. Sahiner, A., Yilmaz, N., Kapusuz, G.: A descent global optimization method based on smoothing techniques via Bezier curves. Carpathian J. Math. **33**(3), 373–380 (2017)Ŕ
23. Sahiner, A., Ibrahem, S.: A new global optimization technique by auxiliary function method in a directional search. Optim. Lett. (2018)
24. Kong, M., Zhuang, J.: A modified filled function method for finding a global minimizer of a nonsmooth function of several variables. Num. Math. J. Chinese Univ. **2**, 165–174 (1996)Ŕ
25. Wu, Z., Bai, F., Lee, H., Yang, Y.: A filled function method for constrained global optimization. J. Global Optim. **39**, 495–507 (2007)
26. Ge, R., Qin, Y.: The globally convexized filled functions for global optimization. Appl. Math. Comput. **35**, 131–158 (1990)
27. Shang, Y., Zhang, L.: A filled function method for finding a global minimizer on global integer optimization. J. Comput. Appl. Math. **181**, 200–210 (2005)
28. Liu, X.: A computable filled function used for global minimization. Appl. Math. Comput. **126**, 271–278 (2002)
29. Wang, W., Shang, Y., Zhang, L.: A filled function method with one parameter for box constrained global optimization. Appl. Math. Comput. **194**, 54–66 (2007)

30. Lin, Y., Yang, Y., Mammadov, M.: A new filled function method for nonlinear equations. Appl. Math. Comput. **210**, 411–421 (2009)
31. Datta, S.: Nanoscale device modeling: the Green's function method. Superlattice Microst. **28**(4), 253–278 (2000)Ŕ
32. Bai, F., Wu, Z., Yang, Y.: A filled function method for constrained nonlinear integer programming. J. Ind. Manag. Optim. **4**, 353–362 (2008)
33. Wang, C., Luo, R., Wu, K., Han, B.: A new filled function method for an unconstrained nonlinear equation. J. Comput. Appl. Math. **235**, 1689–1699 (2011)
34. Wang, X., Yang, G., Yang, Y.: License plate fault-tolerant characters recognition algorithm based on color segmentation and BP neural network. Appl. Mech. Mater. **411–414**, 1281–1286 (2013)
35. Zhang, Y., Xu, Y., Zhang, L.: A filled function method applied to nonsmooth constrained global optimization. J. Comput. Appl. Math. **232**, 415–426 (2009)
36. Liu, X., Xu, W.: A new filled function applied to global optimization. Comput. Oper. Res. **31**, 61–80 (2004)
37. Lin, H., Gao, Y., Wang, Y.: A continuously differentiable filled function method for global optimization. Numer. Algorithms **66**, 511–523 (2013)
38. Lee, J., Lee, B.: A global optimization algorithm based on the new filled function method and the genetic algorithm. Eng. Optim. **27**, 1–20 (1996)
39. Lin, H., Wang, Y., Fan, L.: A filled function method with one parameter for unconstrained global optimization. Appl. Math. Comput. **218**, 3776–3785 (2011)
40. Wei, F., Wang, Y., Lin, H.: A new filled function method with two parameters for global optimization. J. Optim. Theory App. **163**, 510–527 (2014)
41. Zheng, Y.: A class of filled functions for finding a global minimizer of a function of several variables. In: Chinese Science Abstracts Series A 2, (1995)
42. Marler, R., Arora, J.: Survey of multi-objective optimization methods for engineering. Struct. Multidiscip. Optim. **26**, 369–395 (2004)
43. Cooper, V., Nguyen, V., Nicell, J.: Evaluation of global optimization methods for conceptual rainfall-runoff model calibration. Water Sci. Technol. **36**, 53–60 (1997)
44. Gottvald, A., Preis, K., Magele, C., Biro, O., Savini, A.: Global optimization methods for computational electromagnetics. IEEE T. Magn. **28**, 1537–1540 (1992)
45. Renders, J., Flasse, S.: Hybrid methods using genetic algorithms for global optimization. IEEE Trans. Syst. Man. Cybern. B. **26**, 243–258 (1996)
46. Younis, A., Dong, Z.: Trends, features, and tests of common and recently introduced global optimization methods. Eng. Optimiz. **42**, 691–718 (2010)
47. Gashry, G.: On globally convergent multi-objective optimization. Appl. Math. Comput. **183**, 209–216 (2006)
48. Ratschek, H., Rokne, J.: *New Computer Methods for Global Optimization*. Halsted Press, Chichester, Great Britain, (1988)
49. Cheng, Y., Li, L., Chi, S.: Performance studies on six heuristic global optimization methods in the location of critical slip surface. Comput. Geotech. **34**, 462–484 (2007)

Chapter 6
A Modified Laguerre Matrix Approach for Burgers–Fisher Type Nonlinear Equations

Burcu Gürbüz and Mehmet Sezer

6.1 Introduction

Nonlinear partial differential equations play an important role in the study of nonlinear physical phenomena [5–15, 17–22, 24–29]. Especially, Burgers-Fisher and related equations play a major role in various field of science such as biology, chemistry, economics, applied mathematics, physics, engineering, and so on. Burgers–Fisher equation is an important model of fluid dynamics which is of high importance for describing different mechanisms such as convection effect, diffusion transport or interaction between the reaction mechanisms [9].

Specifically, Burgers–Fisher equation considers the effects of nonlinear advection, linear diffusion and nonlinear logistic reaction which describes a nonlinear parabolic mathematical model of various phenomena [23, 41]. This type of equation is a highly nonlinear equation which describes the combination of reaction, convection and diffusion mechanisms [34]. This equation is called as Burgers–Fisher since it combines the properties of convective phenomenon from Burgers equation and diffusion transport as well as reactions related to the characteristics from Fisher equation [2]. In this regard Burgers equation is useful since it models the convective and diffusive terms in the physical applications such as a fluid flow nature in which either shocks or viscous dissipation. It can be used as a

B. Gürbüz (✉)
Üsküdar University, Faculty of Engineering and Natural Sciences, Department of Computer Engineering, Istanbul, Turkey

Johannes Gutenberg University Mainz, Institute of Mathematics, Mainz, Germany
e-mail: burcu.gurbuz@uskudar.edu.tr; burcu.gurbuz@uni-mainz.de

M. Sezer
Manisa Celal Bayar University, Faculty of Science and Letters, Department of Mathematics, Manisa, Turkey
e-mail: mehmet.sezer@cbu.edu.tr

© Springer Nature Switzerland AG 2020

J. A. T. Machado et al. (eds.), *Numerical Solutions of Realistic Nonlinear Phenomena*, Nonlinear Systems and Complexity 31, https://doi.org/10.1007/978-3-030-37141-8_6

model for any nonlinear wave propagation problem subject to dissipation [15, 17–22, 24–30]. It was particularly stressed by Burgers (1948) as being the simplest one to combine typical nonlinearity with typical heat diffusion. Accordingly, it is usually referred to as Burgers equation [7]. Burgers equation is used to demonstrate various computational algorithms for convection-dominated flows [40]. Fisher equation was proposed by Fisher in 1937 which is another very important nonlinear diffusion equation [36]. The Fisher equation is evolution equation that describes the propagation of a virile mutant in an infinitely long habitat [14]. It also represents a model equation for the evolution of a neutron population in a nuclear reactor [8] and a prototype model for a spreading flame [26–33, 35–37].

The study of Burgers–Fisher type nonlinear equations have been considered by many authors for conceptual understanding of physical flows and testing various numerical methods [9]. Mickens gave a nonstandard finite difference scheme for the Burgers–Fisher equation [28], A. Van Niekerk and F. D. Van Niekerk applied Galerkin methods to the nonlinear Burgers equation and obtained implicit and explicit algorithms using different higher order rational basis functions [35], Zhu and Kang introduced the cubic B-spline quasi-interpolation for the generalized Burgers–Fisher equation [46], Al-Rozbayani and Al-Hayalie investigated numerical solution of Burgers–Fisher equation in one dimension using finite differences methods [2], Ismail, Raslan and Rabboh gave Adomian decomposition method for Burgers–Huxley and Burgers–Fisher equations [25]. Also, Ismail and Rabboh presented a restrictive Padé approximation for the solution of the generalized Fisher and Burgers–Fisher equations [24], Babolian and Saeidian [4], Wazwaz gave the analytic approaches for Burgers, Fisher and Huxley equations [39], Chen presented a finite difference method for Burgers–Fisher equation [10], Kheiri introduced an application of the $\frac{G'}{G}$-expansion method for the Burgers, Fisher and Burgers–Fisher equations [26], Zhang presented exact finite difference scheme and nonstandard finite difference scheme for Burgers and Burgers–Fisher equations [45]. Further, Rashidi gave explicit analytical solutions of the generalized Burgers and Burgers–Fisher equations by homotopy perturbation method [31].

In this work, we introduce a numerical scheme to solve Burgers–Fisher type nonlinear equations using the modified Laguerre matrix method. Firstly, equations and given conditions with respect to the collocation points are reduced to a system of algebraic equations with Laguerre coefficients by putting them in the matrix forms. To obtain the modified Laguerre polynomial solution of Burgers–Fisher equations, we use the truncated Laguerre series. After considering the approximate solutions we have the error analysis in order to show the accuracy and to have better approximation, then the results are demonstrated by tables and figures.

The present study proceeds as follows: in Sect. 6.2, model description is presented with details. In Sect. 6.3, we introduce the novel method used to solve the Burgers–Fisher type nonlinear equations in one dimension. Error analysis is given in Sect. 6.4. In the next Sect. 6.5, we apply a computerized approach for finding the numerical solutions of the Burgers, Fisher and Burgers–Fisher equations. Finally, some discussions and conclusions are given in Sect. 6.6.

6.2 The Model

In this section, we introduce the model description of the Burgers–Fisher type nonlinear equations. We deal with the different versions of the Burgers–Fisher type nonlinear equations: singular perturbed generalized Burgers–Fisher equation, Burgers–Fisher equation, Burgers equation and Fisher equation, respectively.

6.2.1 Singular Perturbed Generalized Burgers–Fisher Equation

Singular perturbed problems occur frequently in various branches of applied science and engineering; for example, fluid dynamics, aero dynamics, oceanography, quantum mechanics, chemical reactor theory, reaction-diffusion processes, radiating flows, etc. [33].

Here, we consider the time dependent singularly perturbed generalized Burgers–Fisher initial-boundary value problem given by

$$u_t + \alpha u^{\delta} u_x = \varepsilon u_{xx} + \beta u(1 - u^{\delta}), \;\; 0 \le a \le x \le b, \; t \ge 0, \tag{6.1}$$

with initial condition

$$u(x, 0) = \phi(x), \; x \in \Omega = (0 \le x \le b) \tag{6.2}$$

and boundary conditions as

$$\begin{aligned} u(0, t) &= f(t), \; t \ge 0 \\ u(l, t) &= g(t), \; t \ge 0 \end{aligned} \tag{6.3}$$

over a domain $D = \Omega \times T = (0 \le x \le b) \times t \ge 0$, where α, β and δ are parameters such that $\alpha, \beta \ge 0, \; \delta > 0$ and ε is the singular perturbation parameter $(0 < \varepsilon \ll 1)$.

6.2.2 Burgers–Fisher Equation

The Burgers–Fisher equation has a wide range of applications in plasma physics, fluid dynamics, capillary-gravity waves, nonlinear optics and chemical physics [11]. The Burgers–Fisher equation can be described by Eq. (6.1) for $\delta = 1$. In this case, u represents travelling wave phenomena, u_{xx} corresponds to diffusion term, ε is the diffusion coefficient, α, β and ε are parameters satisfying $\alpha, \beta \ge 0, \; \varepsilon > 0$ and $0 < \varepsilon \le 1$. Keeping other parameters fixed, when $\varepsilon \to 0$, Eq. (6.1) becomes singularly perturbed problem [33]. Here, in particular, we consider Eq. (6.1) for $\delta = 1$ and $\varepsilon = 1$. Then Eq. (6.1) reduces to the Burgers–Fisher equation:

$$u_t + \alpha u u_x = u_{xx} + \beta u(1 - u), \;\; 0 \le a \le x \le b, \; t \ge 0. \tag{6.4}$$

6.2.3 Burgers Equation

Burger, in 1948, placed Burgers equation as significant model of one-dimensional turbulence which defines nonlinear acoustics waves in gases. Moreover, it describes the spread of sound wave and heat exchange in the medium with viscidity. Besides, Burgers equation with variable coefficient describes the cylinder and spherical wave in these questions such as overfall, traffic flow model and so on [11, 12, 14, 15, 17–22, 24–33]. The Burgers equation is a particular case of Eq. (6.1) for $\delta = 1,\ \varepsilon = 1$ and $\beta = 0$, i.e.,

$$u_t + \alpha u u_x = u_{xx},\ \ 0 \le a \le x \le b,\ t \ge 0. \tag{6.5}$$

6.2.4 Fisher Equation

Fisher, in 1937, proposed in his paper the Fisher equation which describes spatial spread of an advantageous allele [33]. Besides having $\delta = 1$ and $\varepsilon = 1$ in Eq. (6.1), we also have $\alpha = 0$. Then we reduce Eq. (6.1) to the Fisher equation, i.e.,

$$u_t = u_{xx} + \beta u(1-u),\ \ 0 \le a \le x \le b,\ t \ge 0. \tag{6.6}$$

6.3 The Method

In this work, the modified Laguerre matrix-collocation method is presented and applied on singular perturbed generalized Burgers–Fisher equation, Burgers–Fisher equation, Burgers equation and Fisher equation, respectively. The method is based on truncated Laguerre series

$$u(x,t) = \sum_{n=0}^{N}\sum_{m=0}^{N} a_{n,m} L_{n,m}(x,t), \tag{6.7}$$

$$L_{n,m}(x,t) = L_n(x)L_m(t),\ \ 0 \le a \le x, t \le b < \infty,$$

where $a_n, n = 0, 1, \ldots, N$ are unknown coefficients to be found and $L_n(x)$ and $L_m(t)$ are the Laguerre polynomials [1–5, 7–12]. We define Laguerre polynomials as

$$L_n(x) = \sum_{k=0}^{n} \frac{(-1)^k}{k!}\binom{n}{k}x^k,\ \ L_m(t) = \sum_{k=0}^{m} \frac{(-1)^k}{k!}\binom{n}{k}t^k. \tag{6.8}$$

and the collocation points are defined by

$$x_i = \frac{l}{N}i,\ t_j = \frac{T}{N}j,\ i, j = 0, 1, 2, \ldots, N. \tag{6.9}$$

Now, let us define the matrix form of Eq. (6.1) by using the matrix form of the solution function (6.7)

$$[u(x,t)] = \mathbf{L}(x)\overline{\mathbf{L}}(t)\mathbf{A}, \tag{6.10}$$

where

$$\mathbf{L}(x) = \begin{bmatrix} L_0(x) & L_1(x) & \cdots & L_N(x) \end{bmatrix},$$

$$\overline{\mathbf{L}}(t) = \begin{bmatrix} \mathbf{L}(t) & 0 & \cdots & 0 \\ 0 & \mathbf{L}(t) & \cdots & 0 \\ \vdots & \vdots & \ddots & \vdots \\ 0 & 0 & \cdots & \mathbf{L}(t) \end{bmatrix},$$

$$\mathbf{A} = \begin{bmatrix} a_{00} & a_{01} & \cdots & a_{0N} & \cdots & a_{N0} & a_{N1} & \cdots & a_{NN} \end{bmatrix}^T.$$

Then we use the relation (6.8) and we define matrix forms of $\mathbf{L}(x)$ and $\overline{\mathbf{L}}(t)$ as [19–21].

$$\mathbf{L}(x) = \mathbf{X}(x)\mathbf{H}$$

$$\overline{\mathbf{L}}(t) = \overline{\mathbf{X}}(t)\overline{\mathbf{H}}, \tag{6.11}$$

where

$$\mathbf{X}(x) = \begin{bmatrix} 1 & x^1 & \cdots & x^N \end{bmatrix},\ \mathbf{X}(t) = \begin{bmatrix} 1 & t^1 & \cdots & t^N \end{bmatrix},$$

$$\mathbf{H} = \begin{bmatrix} \frac{(-1)^0}{0!}\binom{0}{0} & \frac{(-1)^0}{0!}\binom{1}{0} & \frac{(-1)^0}{0!}\binom{2}{0} & \cdots & \frac{(-1)^0}{0!}\binom{N}{0} \\ 0 & \frac{(-1)^1}{1!}\binom{1}{1} & \frac{(-1)^1}{1!}\binom{2}{1} & \cdots & \frac{(-1)^1}{1!}\binom{N}{1} \\ 0 & 0 & \frac{(-1)^2}{2!}\binom{2}{2} & \cdots & \frac{(-1)^2}{2!}\binom{N}{2} \\ \vdots & \vdots & \ddots & \vdots & \vdots \\ 0 & 0 & 0 & \cdots & \frac{(-1)^N}{N!}\binom{N}{N} \end{bmatrix},$$

$$\overline{\mathbf{X}}(t) = diag\left[\mathbf{X}(t)\ \mathbf{X}(t)\ \cdots\ \mathbf{X}(t)\right],$$
$$\overline{\mathbf{H}} = diag\left[\mathbf{H}\ \mathbf{H}\ \cdots\ \mathbf{H}\right].$$

We also show the relation between the matrix $\mathbf{X}(x)$ and its derivatives as

$$\mathbf{X}'(x) = \mathbf{X}(x)\mathbf{B},$$
$$\mathbf{X}''(x) = \mathbf{X}(x)\mathbf{B}^2,$$

where

$$\mathbf{B} = \begin{bmatrix} 0 & 1 & 0 & \cdots & 0 \\ 0 & 0 & 2 & \cdots & 0 \\ \vdots & \vdots & \vdots & \ddots & \vdots \\ 0 & 0 & 0 & \cdots & N \\ 0 & 0 & 0 & \cdots & 0 \end{bmatrix}.$$

Also, by using the matrix relations

$$\mathbf{L}(x) = \mathbf{X}(x)\mathbf{H} \Rightarrow \mathbf{X}(x) = \mathbf{L}(x)\mathbf{H}^{-1}$$

$$\mathbf{L}'(x) = \mathbf{X}'(x)\mathbf{H} = \mathbf{X}(x)\mathbf{B}\mathbf{H}, \tag{6.12}$$

we have

$$\mathbf{L}'(x) = \mathbf{L}(x)\mathbf{C}$$

$$\overline{\mathbf{L}}'(t) = \overline{\mathbf{L}}(t)\overline{\mathbf{C}}, \tag{6.13}$$

where

$$\mathbf{C} = \mathbf{H}^{-1}\mathbf{B}\mathbf{H} = \begin{bmatrix} 0 & -1 & -1 & \cdots & -1 \\ 0 & 0 & -1 & \cdots & -1 \\ \vdots & \vdots & \vdots & \ddots & \vdots \\ 0 & 0 & 0 & \cdots & -1 \\ 0 & 0 & 0 & \cdots & 0 \end{bmatrix};$$

$$\overline{\mathbf{C}} = diag[\mathbf{C}\mathbf{C}\ldots\mathbf{C}].$$

We also have the following equations by using (6.11)–(6.13):

$$\mathbf{L}''(x) = \mathbf{L}'(x)\mathbf{C} = \mathbf{L}(x)\mathbf{C}^2$$

$$\overline{\mathbf{L}}''(t) = \overline{\mathbf{L}}'(t)\overline{\mathbf{C}} = \overline{\mathbf{L}}(t)\overline{\mathbf{C}}^2. \tag{6.14}$$

Therefore, from the relations (6.10)–(6.14):

$$[u(x,t)] = \mathbf{L}(x)\overline{\mathbf{L}}(t)\mathbf{A}$$

$$[u_x(x,t)] = \mathbf{L}'(x)\overline{\mathbf{L}}(t)\mathbf{A} = \mathbf{L}(x)\mathbf{C}\overline{\mathbf{L}}(t)\mathbf{A} \tag{6.15}$$

$$[u_{xx}(x,t)] = \mathbf{L}''(x)\overline{\mathbf{L}}(t)\mathbf{A} = \mathbf{L}(x)\mathbf{C}^2\overline{\mathbf{L}}(t)\mathbf{A} \tag{6.16}$$

$$[u_t(x,t)] = \mathbf{L}(x)\overline{\mathbf{L}}'(t)\mathbf{A} = \mathbf{L}(x)\overline{\mathbf{L}}(t)\overline{\mathbf{C}}\mathbf{A}. \tag{6.17}$$

On the other hand, we also have

$$[u^2(x,t)] = \mathbf{L}(x)\overline{\mathbf{L}}(t)\overline{\overline{\mathbf{L}}}(x)\overline{\overline{\overline{\mathbf{L}}}}(t)\overline{\mathbf{A}} \tag{6.18}$$

$$[u^3(x,t)] = \mathbf{L}(x)\overline{\mathbf{L}}(t)\overline{\overline{\mathbf{L}}}(x)\overline{\overline{\overline{\mathbf{L}}}}(t)\overline{\overline{\overline{\overline{\mathbf{L}}}}}(x)\overline{\overline{\overline{\overline{\overline{\mathbf{L}}}}}}(t)\overline{\overline{\mathbf{A}}}, \tag{6.19}$$

where

$$\begin{aligned}
\mathbf{A}_i &= \begin{bmatrix} a_{i0}\ a_{i1} \cdots a_{iN} \end{bmatrix}^T,\ i = 0, 1, \ldots, N, \\
\mathbf{A} &= \begin{bmatrix} \mathbf{A}_0\ \mathbf{A}_1 \cdots \mathbf{A}_N \end{bmatrix}^T \\
&= \begin{bmatrix} a_{00} \cdots a_{0N}\ a_{10} \cdots a_{1N} \cdots a_{N0} \cdots a_{NN} \end{bmatrix}^T, \\
\overline{\mathbf{A}}_i &= \begin{bmatrix} a_{i0}\mathbf{A}\ a_{i1}\mathbf{A} \cdots a_{iN}\mathbf{A} \end{bmatrix}^T,\ i = 0, 1, \ldots, N, \\
\overline{\mathbf{A}} &= \begin{bmatrix} \overline{\mathbf{A}}_0\ \overline{\mathbf{A}}_1 \cdots \overline{\mathbf{A}}_N \end{bmatrix}^T, \\
\overline{\overline{\mathbf{A}}}_i &= \begin{bmatrix} a_{i0}\overline{\mathbf{A}}\ a_{i1}\overline{\mathbf{A}} \cdots a_{iN}\overline{\mathbf{A}} \end{bmatrix}^T,\ i = 0, 1, \ldots, N, \\
\overline{\overline{\mathbf{A}}} &= \begin{bmatrix} \overline{\overline{\mathbf{A}}}_0\ \overline{\overline{\mathbf{A}}}_1 \cdots \overline{\overline{\mathbf{A}}}_N \end{bmatrix}^T.
\end{aligned}$$

Furthermore,

$$[u_x(x,t)u(x,t)] = \mathbf{L}(x)\ mathbf C\overline{\mathbf{L}}(t)\overline{\overline{\mathbf{L}}}(x)\overline{\overline{\overline{\mathbf{L}}}}(t)\overline{\mathbf{A}} \tag{6.20}$$

$$[u_x(x,t)u^2(x,t)] = \mathbf{L}(x)\mathbf{C}\overline{\mathbf{L}}(t)\overline{\overline{\mathbf{L}}}(x)\overline{\overline{\overline{\mathbf{L}}}}(t)\overline{\overline{\overline{\overline{\mathbf{L}}}}}(x)\overline{\overline{\overline{\overline{\overline{\mathbf{L}}}}}}(t)\overline{\overline{\mathbf{A}}}. \tag{6.21}$$

Now, we organize Eq. (6.1) as

$$u_t + \alpha u_x u^\delta - \varepsilon u_{xx} - \beta u + \beta u^{\delta+1} = 0,\ \ 0 \le a \le x \le b,\ t \ge 0. \tag{6.22}$$

Then we replace the matrix relations (6.10), (6.16) and (6.17)–(6.21) into Eq. (6.22) for $\delta = 2$ and we have

$$
\begin{aligned}
&\{\mathbf{L}(x)\overline{\mathbf{L}}(t)\overline{\mathbf{C}}-\varepsilon\mathbf{L}(x)\mathbf{C}^2\overline{\mathbf{L}}(t) - \beta\mathbf{L}(x)\overline{\mathbf{L}}(t)\}\mathbf{A} \\
&\quad+\{\alpha\mathbf{L}(x)\mathbf{C}\overline{\mathbf{L}}(t)\overline{\overline{\mathbf{L}}}(x)\overline{\overline{\overline{\mathbf{L}}}}(t)\overline{\overline{\overline{\overline{\mathbf{L}}}}}(x)\overline{\overline{\overline{\overline{\overline{\mathbf{L}}}}}}(t) \\
&\quad+\beta\mathbf{L}(x)\overline{\mathbf{L}}(t)\overline{\overline{\mathbf{L}}}(x)\overline{\overline{\overline{\mathbf{L}}}}(t)\overline{\overline{\overline{\overline{\mathbf{L}}}}}(x)\overline{\overline{\overline{\overline{\overline{\mathbf{L}}}}}}(t)\}\overline{\overline{\mathbf{A}}} = [0].
\end{aligned}
\tag{6.23}
$$

For $\delta = 1$, we obtain

$$
\begin{aligned}
&\{\mathbf{L}(x)\overline{\mathbf{L}}(t)\overline{\mathbf{C}}-\varepsilon\mathbf{L}(x)\mathbf{C}^2\overline{\mathbf{L}}(t) - \beta\mathbf{L}(x)\overline{\mathbf{L}}(t)\}\mathbf{A} \\
&\quad+\{\alpha\mathbf{L}(x)\mathbf{C}\overline{\mathbf{L}}(t)\overline{\overline{\mathbf{L}}}(x)\overline{\overline{\overline{\mathbf{L}}}}(t) + \beta\mathbf{L}(x)\overline{\mathbf{L}}(t)\overline{\overline{\mathbf{L}}}(x)\overline{\overline{\overline{\mathbf{L}}}}(t)\}\overline{\mathbf{A}} = [0].
\end{aligned}
\tag{6.24}
$$

From (6.23) and (6.24), concisely, we have

$$\mathbf{W}(x,t)\mathbf{A} + \mathbf{W}^*(x,t)\overline{\overline{\mathbf{A}}} = [0], \tag{6.25}$$

$$\mathbf{W}(x,t)\mathbf{A} + \mathbf{W}^{**}(x,t)\overline{\mathbf{A}} = [0]. \tag{6.26}$$

Similarly, we use the procedure for initial and boundary conditions in (6.2)–(6.3) for $i, j = 0, 1, \ldots, N$:

$$
\begin{aligned}
[u(x,0)] &= \mathbf{L}(x)\overline{\mathbf{L}}(0)\mathbf{A} = [\phi(x)] = \lambda \\
[u(0,t)] &= \mathbf{L}(0)\overline{\mathbf{L}}(t)\mathbf{A} = [f(t)] = \mu \\
[u(l,t)] &= \mathbf{L}(l)\overline{\mathbf{L}}(t)\mathbf{A} = [g(t)] = \gamma.
\end{aligned}
\tag{6.27}
$$

Now, we consider the collocation points in Eq. (6.9) and replace in Eq. (6.25) and in Eq. (6.26):

$$
\begin{aligned}
\mathbf{W}(x_i,t_j)\mathbf{A} + \mathbf{W}^*(x_i,t_j)\overline{\overline{\mathbf{A}}} &= \mathbf{0}, \\
\mathbf{W}(x_i,t_j)\mathbf{A} + \mathbf{W}^{**}(x_i,t_j)\overline{\mathbf{A}} &= \mathbf{0}.
\end{aligned}
$$

Briefly, the fundamental matrix equations can be defined as

$$\mathbf{WA} + \mathbf{W}^*\overline{\overline{\mathbf{A}}} = \mathbf{0} \Longrightarrow [\mathbf{W}; \mathbf{W}^* : \mathbf{0}], \tag{6.28}$$

$$\mathbf{WA} + \mathbf{W}^{**}\overline{\mathbf{A}} = \mathbf{0} \Longrightarrow [\mathbf{W}; \mathbf{W}^{**} : \mathbf{0}]. \tag{6.29}$$

Also, we use the collocation points in Eq. (6.9) and replace in Eq. (6.27):

$$\begin{aligned}
[u(x_i, 0)] &= \mathbf{L}(x_i)\overline{\mathbf{L}}(0)\mathbf{A} = [\phi(x_i)] = \lambda_i \\
[u(0, t_j)] &= \mathbf{L}(0)\overline{\mathbf{L}}(t_j)\mathbf{A} = [f(t_j)] = \mu_j \\
[u(l, t_j)] &= \mathbf{L}(l)\overline{\mathbf{L}}(t_j)\mathbf{A} = [g(t_j)] = \gamma_j
\end{aligned} \tag{6.30}$$

or in the short form:

$$\begin{aligned}
\mathbf{UA} &= [\lambda];\ \ [\mathbf{U} : \lambda], \\
\mathbf{VA} &= [\mu];\ \ [\mathbf{V} : \mu], \\
\mathbf{ZA} &= [\gamma];\ \ [\mathbf{Z} : \gamma].
\end{aligned}$$

Then we organize the augmented matrix in the new form by removing the last rows of (6.28) and (6.29). So, we get

$$\begin{aligned}
&[\ \tilde{\mathbf{W}}; \tilde{\mathbf{W}^{*}} : \tilde{\mathbf{0}}], \\
&[\ \tilde{\mathbf{W}}; \tilde{\mathbf{W}^{**}} : \tilde{\mathbf{0}}].
\end{aligned}$$

Lately, the augmented matrix form systems are solved by Gaussian elimination. Then the unknown Laguerre coefficients are computed [22]. Thus, the approximate solutions, $u(x, t),\ 0 \leq a \leq x \leq b,\ t \geq 0$, for the singular perturbed generalized Burgers–Fisher and Burgers–Fisher equations are found in the truncated Laguerre series form as following:

$$u(x, t) \cong u_N(x, t) = \sum_{n=0}^{N} \sum_{m=0}^{N} a_{n,m} L_{n,m}(x, t). \tag{6.31}$$

In a similar way, we can generalize the method by having $\delta = 1, 2, \ldots, n$. Also, in order to find the approximate solution of Burgers equation for $\delta = 1,\ \varepsilon = 1$ and $\beta = 0$, Eq. (6.5), we repeat the process. Correspondingly, the approximate solution of the Fisher equation, Eq. (6.6), can be found by following the same procedure for $\delta = 1$ and $\varepsilon = 1$.

6.4 Error Analysis

In this section, the error estimation for the Laguerre polynomial solution (6.7) is given which shows the accuracy of the method. We define error function $x = x_\zeta, t = t_\eta \in [l, 0] \times [0, T],\ \zeta, \eta = 0, 1, \ldots$

$$E_N(x_p, t_q) = |\ u(x_p, t_q) - (u_N)_t(x_p, t_q) - \alpha(u_N)^\delta(x_p, t_q)(u_N)_x(x_p, t_q)$$
$$+ \varepsilon(u_N)_{xx}(x_p, t_q) + \beta(u_N)(x_p, t_q)(1 - (u_N)^\delta(x_p, t_q))\ | \cong 0,$$

where $E_N(x_p, t_q) \leq 10^{(-k_{\zeta\eta})} = 10^{(-k)}$, ($k$ is a positive integer) is prescribed, then the truncation limit N increased until difference $E_N(x_p, t_q)$ at each of the points becomes smaller than the prescribed $10^{(-k)}$. Furthermore, we measure errors with respect to different type of error norms which are defined as follows:

1. For L_2; $E_N(x_p, t_q) = (\sum_{i=1}^{n}(e_i)^2)^{1/2}$
2. For L_∞; $E_N(x_p, t_q) = Max(e_i),\ 0 \leq i \leq n$
3. For RMS; $E_N(x_p, t_q) = \sqrt{\frac{\sum_{i=1}^{n+1}(e_i)^2}{n+1}}$,

where RMS is the Root-Mean-Square of errors and $e_i = u(x_i, \tau) - \check{u}(x_i, \tau)$; also u is the exact and $\check{u}$ is the approximate solutions of the problem. Also, τ and t are arbitrary time variables in $[0, T]$ [1].

6.4.1 Residual Error Estimation

Here, we consider the Burgers–Fisher equation to show its residual error estimation technique based on modified Laguerre collocation method [18]. We consider Eq. (6.1) related to operator L and we have

$$L[u(x, t)] = u_t + \alpha u^\delta u_x - \varepsilon u_{xx} - \beta u(1 - u^\delta) = 0,\quad 0 \leq a \leq x \leq b,\ t \geq 0.$$

Then, we define the residual function of the modified Laguerre collocation method as

$$R_N(x, t) = L[u_N(x, t)],$$

where $u_N(x, t)$ is the approximate solution obtained by modified Laguerre collocation method with conditions (6.2)–(6.3). Hence, this satisfies the problem

$$(u_N)_t + \alpha(u_N)^\delta(u_N)_x - \varepsilon(u_N)_{xx} - \beta(u_N)(1-(u_N)^\delta) = R_N(x, t),$$
$$0 \leq a \leq x \leq b,\ t \geq 0,$$

with the conditions

$$u_N(x, 0) = \phi(x),\ x \in \Omega = (0 \leq x \leq b)$$
$$u_N(0, t) = f(t),\ t \geq 0$$
$$u_N(l, t) = g(t),\ t \geq 0.$$

We have defined the error function as $e_N = u(x,t) - u_N(x,t)$ with the homogenous conditions

$$L[e_N(x,t)] = L[u(x,t)] - L[u_N(x,t)] = -R_N(x,t)$$

then the error problem can be defined as

$$(e_N)_t + \alpha(e_N)^{\delta}(e_N)_x - \varepsilon(e_N)_{xx} - \beta(e_N)(1-(e_N)^{\delta}) = -R_N(x,t),$$
$$0 \le \quad a \le x \le b,\ t \ge 0.$$

$$e_N(x,0) = 0,\ x \in \Omega = (0 \le x \le b)$$
$$e_N(0,t) = 0,\ t \ge 0$$
$$e_N(l,t) = 0,\ t \ge 0.$$

By solving this problem with the introduced technique, we find the approximation to $e_N(x,t)$ by

$$e_{N,M}(x,t) \cong \sum_{n=0}^{N}\sum_{m=0}^{N} a^*_{n,m} L_{n,m}(x,t),\quad M > N.$$

Also, we have the improved Laguerre polynomial solution as

$$u^*_{N,M}(x,t) = u_N(x,t) + e_{N,M}(x,t).$$

Thereby, the corrected error function is

$$e^*_{N,M}(x,t) = e_N(x,t) - e_{N,M}(x,t) = u(x,t) - u_{N,M}(x,t)$$

where $e_{N,M}(x,t)$ is the estimated error function [43].

Algorithm

- **Step 0.** Input data: $\beta(t)$ and $\alpha(t)$. Determine the initial and boundary conditions.
- **Step 1.** Set N where $\mathbb{N}$.
- **Step 2.** Construct the indicated matrices.
- **Step 3.** Define the collocation points $x_i = \frac{l}{N}i,\ t_j = \frac{T}{N}j,\ i,j = 0,1,2,\ldots,N$.
- **Step 4.** Compute $[\mathbf{W};\ \mathbf{W}^* : \mathbf{0}]$.
- **Step 5.** Compute $\mathbf{UA} = [\lambda];\ [\mathbf{U} : \lambda],\ \mathbf{VA} = [\mu];\ [\mathbf{V} : \mu],\ \mathbf{ZA} = [\gamma];\ [\mathbf{Z} : \gamma]$.
- **Step 6.** Construct the augmented matrix $[\tilde{\mathbf{W}};\ \tilde{\mathbf{W}}^* : \tilde{\mathbf{0}}]$
- **Step 7.** Solve the system by Gaussian elimination method. Print $\mathbf{A}$.
- **Step 8.** Replace the elements $a_{m,n}$ from Step 7 in the truncated Laguerre series form.
- **Step 9.** Output data: the approximate solution $u_N(x,t)$.
- **Step 10.** Construct $u(x,t)$ is the exact solution of (1).
- **Step 11.** Stop when $|e_i| \le 10^{-k}$ where $k \in \mathbb{Z}^+$. Otherwise, increase N and return to Step 1 [16, 21].

6.5 Numerical Results

In this section, we introduce numerical examples to show the accuracy of the method.

Example 6.1 Firstly, we consider the generalized singularly perturbed Burgers–Fisher equation in Eq. (6.1) for $\delta = 1$ and also for the specific parameter values of $\alpha = 0.01,\ \beta = 0.01$ together with the initial condition [33]

$$u(x,0) = \frac{1}{2} + \frac{1}{2}\tanh(\theta_1, x),\ 0 \le x \le 1$$

and boundary conditions

$$u(0,t) = \frac{1}{2} + \frac{1}{2}\tanh(0 - \theta_1\theta_2 t),\ t \ge 0$$

$$u(1,t) = \frac{1}{2} + \frac{1}{2}\tanh(\theta_1 - \theta_1\theta_2 t),\ t \ge 0$$

and the analytic solution is given by

$$u(x,t) = \frac{1}{2} + \frac{1}{2}\tanh(\theta_1 x - \theta_1\theta_2 t),$$

where $\theta_1 = -\frac{\alpha}{4\varepsilon}$ and $\theta_2 = \frac{\alpha}{2} + \frac{2\varepsilon\beta}{\alpha}$ [45]. Also, $\varepsilon = 1$ has been chosen and we may see the difference between exact and approximate solutions for $N = 4$ and $N = 6$ in Figs. 6.1, 6.2 and 6.3, respectively. In Table 6.1. the maximum absolute errors of the present method and the other methods such as variational iteration method (VIM), lattice Boltzmann method (LBM) and Adomian decomposition method (ADM) have been shown for $\alpha = \beta = 0.01$, $\varepsilon = 1$, $t = 1$ and $x = 0.1,\ 0.5$ and 0.9.

Example 6.2 Now, we consider the Burgers–Fisher equation in Eq. (6.4) for the specific parameter values of $\alpha = -1,\ \beta = -1$ together with the initial condition

$$u(x,0) = \frac{1}{2} + \frac{1}{2}\tanh\left(\frac{-x}{4}\right),\quad 0 \le x,\ t \le 1$$

is given. Also, the exact solution of the problem is $u(x,t) = \frac{1}{2} + \frac{1}{2}\tanh\left(\frac{x}{4} + \frac{5t}{8}\right)$ [32, 33, 35–38]. In Table 6.2, the comparison has been given between absolute error functions E_2 and E_4 and the estimated absolute errors for $E_{2,4}$ and $E_{4,6}$.

Example 6.3 We simulate Burgers Equation in Eq. (6.5) with the initial-boundary value problem [45], for $\alpha = 1,\ \beta = 1$

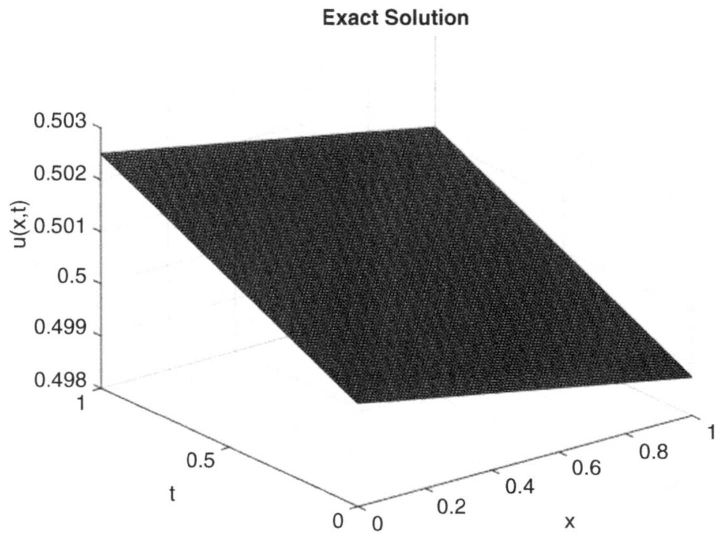

Fig. 6.1 Exact solution for Example 6.1

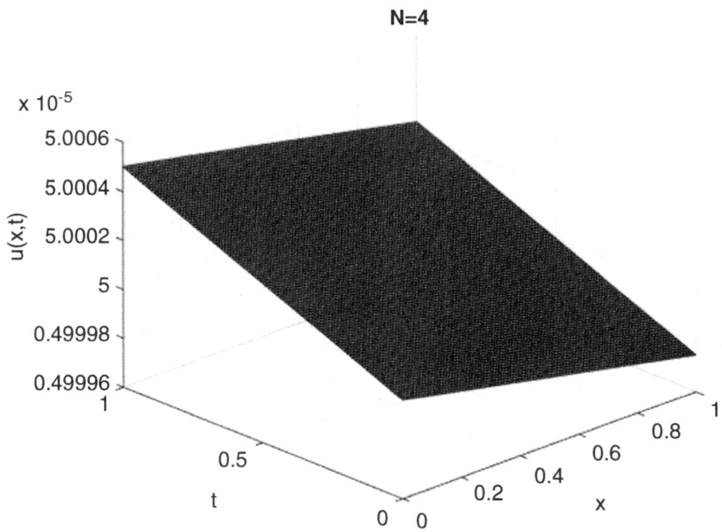

Fig. 6.2 Approximate solution of $N = 4$ for Example 6.1

$$u(x,0) = \frac{1}{1+\exp(\frac{x}{2})}, \quad 0 \le x \le 1$$

$$u(0,t) = \frac{1}{1+\exp(\frac{-t}{4})}, \quad 0 \le t$$

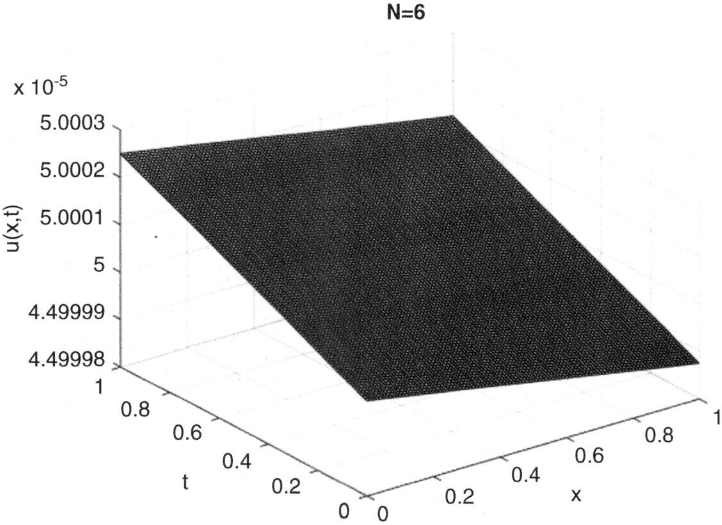

Fig. 6.3 Approximate solution of $N = 6$ for Example 6.1

Table 6.1 Comparison of the maximum absolute errors of Example 1 [33]

t	x	Present scheme	LBM	VIM	ADM
1	0.1	2.007×10^{-6}	1.080×10^{-4}	1.780×10^{-8}	1.780×10^{-8}
	0.5	5.180×10^{-9}	0.325×10^{-4}	5.290×10^{-9}	5.290×10^{-9}
	0.9	4.999×10^{-8}	1.730×10^{-4}	7.280×10^{-9}	7.280×10^{-9}

Table 6.2 Comparison of the actual and estimated absolute errors for $N = 2$ and 4 and $M = 4$ and 6 of Example 6.2

x	t	E_2	$E_{2,4}$	E_4	$E_{4,6}$
0.0	0.0	1.027×10^{-3}	1.001×10^{-3}	4.501×10^{-5}	1.030×10^{-6}
0.1	0.1	1.865×10^{-3}	2.352×10^{-4}	4.602×10^{-5}	1.905×10^{-6}
0.2	0.2	3.095×10^{-3}	7.697×10^{-4}	9.493×10^{-6}	3.662×10^{-6}
0.3	0.3	6.208×10^{-4}	5.862×10^{-4}	7.288×10^{-6}	4.335×10^{-6}
0.4	0.4	3.644×10^{-4}	3.732×10^{-4}	5.162×10^{-6}	3.702×10^{-6}
0.5	0.5	1.159×10^{-4}	1.029×10^{-4}	5.200×10^{-6}	2.906×10^{-6}
0.6	0.6	4.380×10^{-4}	5.905×10^{-4}	4.338×10^{-6}	2.025×10^{-6}
0.7	0.7	4.653×10^{-5}	5.050×10^{-5}	3.451×10^{-7}	2.771×10^{-7}
0.8	0.8	6.849×10^{-5}	3.037×10^{-5}	7.050×10^{-7}	1.906×10^{-7}
0.9	0.9	7.605×10^{-4}	8.402×10^{-5}	6.009×10^{-6}	5.082×10^{-7}
1.0	1.0	3.032×10^{-4}	1.130×10^{-5}	4.011×10^{-6}	6.005×10^{-7}

Table 6.3 L_2, L_∞ and RMS errors for $N = 3$ and $t = 0.2$ of Example 6.3

x	L_2	L_∞	RMS
1	0.7560×10^{-5}	0.5247×10^{-4}	0.1000×10^{-6}
2	0.1164×10^{-5}	0.3791×10^{-4}	0.1502×10^{-5}
3	0.1550×10^{-4}	0.5467×10^{-3}	0.6855×10^{-4}
4	0.8259×10^{-3}	0.7795×10^{-3}	0.1752×10^{-3}
5	0.4643×10^{-4}	0.5467×10^{-2}	0.2916×10^{-5}

Table 6.4 CPU times for $N = 2$ and $N = 4$ of Example 6.3

Wall clock time (s)	
$N = 2$	$N = 4$
195.288	353.98

$$u(1,t) = \frac{1}{1 + \exp(\frac{1}{2} - \frac{t}{4})}, \quad 0 \leq t.$$

In Table 6.3, error comparison between the norms L_2, L_∞ and Root-Mean-Square RMS have been demonstrated for $N = 3$ and $t = 0.2$. Moreover, a central processing unit (CPU) has been shown on the Table 6.4.

6.6 Concluding Remarks

In this study, a modified Laguerre matrix-collocation method has been introduced to solve Burgers–Fisher type nonlinear equations under the initial and boundary conditions. An error estimation has been implemented by using the residual function to improve the numerical solutions. This improvement has been shown on tables. Approximate solutions have been obtained with different truncation limits, N and M values. The implementations of the results can be seen from figures and tables. As we can see the execution on tables and figures, the errors decrease when N and M are increased.

Furthermore, comparison of the numerical results between the present technique and existed methods has been done. Meanwhile, applicability, efficiency and reliability of the present method have been proved by illustrative examples. A remarkable advantage of the method is that the approximate solutions are computed very comfortably by using a well-known symbolic software such as Matlab, Maple and Mathematica. Additionally, the calculation has been explained by the algorithm and a table has been given to show CPU running time for the computer programmes [17–22, 24–33, 35–44].

Consequently, in this study, such easily convenient and improved technique has been analysed. The method can also be applied on many other real-world problems, though some modifications are required [47].

Acknowledgements The authors would like to thank the referees for the helpful suggestions.

References

1. Aizenshtadt, S.V., Krylov I.V., Metel'skii, S.A.: Tables of Laguerre Polynomials and Functions. Pergamon Press, Poland (1966)
2. Al-Rozbayani, A.M., Al-Hayalie, K.A.: Numerical solution of Burger's Fisher equation in one dimensional using finite differences methods. J. Appl. Math. **2014**, 1–12 (2014)
3. Anguelov, R., Lubuma, J.M-S.: Contributions to the mathematics of the nonstandard finite difference method and applications. Numer. Meth. Part. D. E. **17**(5), 518–543 (2001)
4. Babolian, E., Saeidian, J.: Analytic approximate solutions to Burgers, Fisher, Huxley equations and two combined forms of these equations. Commun. Nonlinear Sci. Numer. Simul. **14**, 1984–1992 (2009)
5. Bekir, A., Boz, A.: Exact solutions for nonlinear evolution equations using Exp-function method. Phys. Lett. A. **372**, 1619–1625 (2008)
6. Broy, M.: Software engineering — from auxiliary to key technologies. In: Broy, M., Dener, E. (eds.) Software Pioneers, pp. 10–13. Springer, Heidelberg (2002)
7. Burgers, J.M.: A mathematical model illustrating the theory of turbulence. Adv. Appl. Mech. Elsevier **1**, 171–199 (1948)
8. Canosa, J.: On a nonlinear diffusion equation describing population growth. IBM J. Res. Develop. **17**, 307–313 (1973)
9. Chandraker, V., Awasthi, A., Jayaraj, S.: Numerical treatment of Burger-Fisher equation. Procedia Technol. **25**, 1217–1225 (2016)
10. Chen, X.Y.: Numerical methods for the Burgers-Fisher equation, Master Thesis. University of Aeronautics and Astronautics, China (2007)
11. Chen, B.-K., Li, Y., Chen, H.-L., Wang, B.-H.: Exp-function method for solving the Burgers-Fisher equation with variable coefficients. arXiv preprint arXiv:1004.1815 (2010)
12. Dieudonné, J.: Orthogonal Polynomials and Applications. Berlin, New York (1985)
13. Dod, J.: Effective Substances. In: The Dictionary of Substances and Their Effects. Royal Society of Chemistry (1999). Available via DIALOG. http://www.rsc.org/dose/titleofsubordinatedocument. Cited 15 Jan 1999
14. Fisher, R.A.: The wave of advance of advantageous genes. Ann. Eugenic. **7**, 355–369 (1937)
15. Fletcher, C.A.J.: Computational Techniques for Fluid Dynamics 1: Fundamental and General Techniques, 2nd edn. Springer, Berlin (1991)
16. Geddes, K.O., Czapor, S.R., Labahn, G.: Algorithms for Computer Algebra. Kluwer, Boston (1992)
17. Gökmen, E., Gürbüz, B., Sezer, M.: A numerical technique for solving functional integro-differential equations having variable bounds. Comput. Appl. Math. 1–15 (2018)
18. Gürbüz, B., Sezer, M., Güler, C.: Laguerre collocation method for solving Fredholm integro-differential equations with functional arguments. J. Appl. Math. **2014** (2014)
19. Gürbüz, B., Sezer, M.: Laguerre polynomial approach for solving nonlinear Klein-Gordon equations. Malays. J. Math. Sci. **11**(2), 191–203 (2017)
20. Gürbüz, B., Sezer, M.: Laguerre polynomial solutions of a class of delay partial functional differential equations. Acta Phys. Pol. A. **132**(3), 558–60 (2017)
21. Gürbüz, B., Sezer, M.: Modified Laguerre collocation method for solving 1-dimensional parabolic convection-diffusion problems. Math. Meth. Appl. Sci. 1–7 (2017)
22. Gürbüz, B., Sezer, M.: A numerical solution of parabolic-type Volterra partial integro-differential equations by Laguerre collocation method. Int. J. Appl. Phys. Math. **7**(13), 49–58 (2017). doi: https://doi.org/10.17706/ijapm.2017.7.1.49-58
23. Hamburger, C.: Quasimonotonicity, regularity and duality for nonlinear systems of partial differential equations. Ann. Mat. Pura. Appl. **169**, 321–354 (1995)
24. Ismail, H.N.A., Rabboh, A.A.A.: A restrictive Padé approximation for the solution of the generalized Fisher and Burger-Fisher equation. Appl. Math. Comput. **154**, 203–210 (2004)
25. Ismail, H.N.A., Raslan, K., Rabboh, A.A.A.: Adomian decomposition method for Burgers-Huxley and Burgers-Fisher equations. Appl. Math. Comput. **159**, 291–301 (2004)

26. Kheiri, H., Ebadi, G.: Application of the $\frac{G'}{G}$ expansion method for the Burgers, Fisher and Burgers Fisher equations. Acta Univ. Apulensis Math. Inform. **24**, 35–44 (2010)
27. Kocaçoban, D., Koç, A.B., Kurnaz, A., Keskin, Y.: A better approximation to the solution of Burger-Fisher equation. In: Proceedings of the World Congress on Engineering, vol. 2011(1), (2011)
28. Mickens, R.E., Gumel, A.B.: Construction and analysis of a non-standard finite difference scheme for the Burgers-Fisher equation. J. Sound Vib. **257**(4), 791–797 (2002)
29. Murray, J.D.: Mathematical Biology: I. An Introduction, 3rd edn. Springer, Berlin (2002)
30. Nawaz, R., Ullah, H., Islam, S., Idrees, M.: Application of optimal homotopy asymptotic method to Burger equations. J. Appl. Math. **2013**, 1–8 (2013)
31. Rashidi, M.M., Ganji, D.D., Dinarvand, S.: Explicit analytical solutions of the generalized Burger and Burger-Fisher equations by homotopy perturbation method. Numer. Meth. Part. D. E. **25**(2), 409–417 (2009)
32. Rong-Pei, Z., Li-Wei, Z.: Direct discontinuous Galerkin method for the generalized Burgers-Fisher equation. Chin. Phys. B. **21**, 1–4 (2012). doi: https://doi.org/10.1088/1674-1056/21/9/090206
33. Sangwan, V., Kaur, B.: An exponentially fitted numerical technique for singularly perturbed Burgers-Fisher equation on a layer adapted mesh. Int. J. Comput. Math. **96**(7), 1502–1513 (2019)
34. Slifka, M.K., Whitton, J.L.: Clinical implications of dysregulated cytokine production. J. Mol. Med. (2000). doi: https://doi.org/10.1007/s001090000086
35. Van Niekerk, A., Van Niekerk, F.D.: A Galerkin method with rational basis functions for burgers equation. Comput. Math. Appl. **20**(2), 45–51 (1990)
36. Wang, X., Lu, Y., Exact solutions of the extended Burgers-Fisher equation. Chin. Phys. Lett. **7**(4), 145 (1990)
37. Wang, X.Y., Zhu, Z.S., Lu, Y.K.: Solitary wave solutions of the generalised Burgers-Huxley equation. J. Phys. A. **23**(3), 271–274 (1990)
38. Wang, X., Lu, Y.: Exact solutions of the extended Burgers-Fisher equation. Chin. Phys. Lett. **7**(4), 145–147 (1990)
39. Wazwaz, A.M.: Analytic study on Burgers, Fisher, Huxley equations and combined forms of these equations. Appl. Math. Comput. **195**, 754–761 (2008)
40. Whitham, G.B.: Linear and Nonlinear Waves, Pure and Applied Mathematics: A Wiley Series of Texts, Monographs and Tracts, 1st edn. Wiley, New York, USA (1974)
41. Yadav, O.P., Jiwari, R.: Finite element analysis and approximation of Burgers'-Fisher equation. Numer. Meth. Part. D. E. **33**, 1652–1677 (2017)
42. Yıldızhan, İ., Kürkçü, Ö.K., Sezer, M.: A numerical approach for solving pantograph-type functional differential equations with mixed delays using Dickson polynomials of the second kind. J. Sci. Arts **18**(3), 667–680 (2018)
43. Yüzbaşı, Ş., Sezer, M.: An exponential approximation for solutions of generalized pantograph-delay differential equations. Appl. Math. Model. **37**(22), 9160–9173 (2013)
44. Yüzbaşı, Ş., Kemancı, B., Sezer, M.: Numerical solutions of integro-differential equations and application of a population model with an improved Legendre method. Appl. Math. Model. **37**(2013), 2086–2101 (2013)
45. Zhang, L., Wang, L., Ding, X.: Exact finite difference scheme and nonstandard finite difference scheme for Burgers and Burgers-Fisher equations. J. Appl. Math. **2014**(2014), 1–12 (2014)
46. Zhu, C-G., Kang, W.S.: Numerical solution of Burgers Fisher equation by cubic B-spline quasi-interpolation. Appl. Math. Comput. **216**(2010), 2679–2686 (2010)
47. Zun-Tao, F., Shi-Da, L., Shi-Kuo, L.: Notes on solutions to Burgers-type equations. Commun. Theor. Phys. **41**(4), 527 (2004)

Chapter 7
Increasing the Effects of Auxiliary Function by Multiple Extrema in Global Optimization

Ahmet Sahiner, Shehab A. Ibrahem, and Nurullah Yilmaz

7.1 Introduction

Many real-life problems have been formulated as optimization problems. They have been applied in many branches of real-life such as finance, portfolio selection, medical science, data mining, etc. [1–4]. Global optimization constitutes one important part of the theory of optimization. It has many application areas in engineering such as electrical power distribution and design of space trajectories [5, 6]. Global optimization is a very active research area because of the problems becoming more and more complicated from year to year due to increasing number of variables and structure of the problems (non-smoothness). Up to now, many new theories and algorithms have been presented to solve global optimization problems [7, 8]. There exist two different type of methods which are based on local searches (Monotonic Basin Hopping, Hill Climbing Methods, etc. [9, 10]) and not based on local searches (Branch and Bound, DIRECT, etc. [11, 12]). We consider the methods based on local searches. For local search based methods the major difficulties for global optimization are listed below:

a. When finding any local minimizer by using a local solver, how to escape from this current local minimizer.
b. How to ignore the local minimizers of which their values are greater than the value of current minimizer and find a lower minimizer of the objective function.
c. How to evaluate the convergence to the global minimizer, and determine the stopping criteria.

A. Sahiner (✉) · S. A. Ibrahem · N. Yilmaz
Department of Mathematics, Süleyman Demirel University, Isparta, Turkey
e-mail: ahmetsahiner@sdu.edu.tr; nurullahyilmaz@sdu.edu.tr

© Springer Nature Switzerland AG 2020
J. A. T. Machado et al. (eds.), *Numerical Solutions of Realistic Nonlinear Phenomena*, Nonlinear Systems and Complexity 31, https://doi.org/10.1007/978-3-030-37141-8_7

In different point of view, global optimization approaches can be classified into three main classes: stochastic methods, heuristic methods, and deterministic methods.

Stochastic methods are quite simple, very efficient in black box problems and robust with respect to the increment of dimension of the problem but some of the stochastic methods can find only a local solution instead of the global one. The very well-known stochastic approaches are Random Search and Adaptive Search, Markovian Algorithms, etc. [13, 14]. The population algorithms are included in stochastic methods but we handle them in the heuristic methods.

The heuristic methods are based on the simulation of the biological, physical, or chemical processes. These methods are easy applicable and they converge the solution rapidly. However, they can give different results if they are run again. The very well-known methods are Genetic Algorithm [15], Simulated Annealing Algorithm [16–18], Particle Swarm Optimization [19, 20], and Artificial Bee Colony Algorithm [21, 22]. In recent years, the hybridizations of the heuristic global optimization algorithms have come into prominence [23–27].

The convergence to the solution is guaranteed in deterministic approaches. This is the outstanding property of the deterministic approaches. However, these methods converge the solution quite slowly [28]. There exist important methods such as Branch and Bound algorithms [11], Covering methods [12], Space Filling Curve methods [29, 30], and other methods [31, 32].

Auxiliary function approach is one of the most important one among the methods on global optimization. These methods are developed according to deterministic search strategies by constructing an auxiliary function to escape from the current local minimizer to a better one, among such methods are Tunneling Method [33], Filled Function Method (FFM) [34–36], and Global Descent Method [37].

The first auxiliary function method was introduced by Levy and Montalvo [33]. Cetin et al. developed the tunneling algorithm to resolve constrained global optimization problems [38]. However, many important studies related to tunneling algorithms have been published in [39–41].

Among other methods, FFM can be considered an effective approach to solve different global optimization problems, so it seems to have several features over others, for example, it is more simple to find a better local minimizer sequentially compared to other methods. The FFM was presented for the first time by Ge [34, 35] and improved in [42–44]. Many valuable studies have been presented in order to make filled function applicable for different type of problems such as non-smooth problems [45, 46], constrained optimization problems [47], system of nonlinear equations [48], etc. [49, 50]. Recently, the next generation of filled function or auxiliary function approaches have been developed [51–55].

The FFM presents a good idea for solving global optimization problems. In general, the filled function mechanism is described as follows:

1. Choose any random point for starting and use a convenient local optimization method to find a local minimizer of the objective function.

2. Construct a filled function based on the current minimizer of the objective function, and use any point in the proximity of this current minimizer to minimize the filled function. Finally, a local minimizer of the filled function is obtained. This minimizer is in a basin of better solution of objective function.
3. The minimizer of filled function which obtained in step 2 is used as a starting point to find the minimizer of the objective function.

Surely the number of minimizers is reduced by repeating Step 2 and 3. Finally, the global minimizer of objective function is found.

Some of the existing filled functions have been constructed to have a surface somewhat like a surface of the objective function in the lower basin (when $f(x) \geq f(x_1^*)$, x_1^* is a current minimizer of the objective function) of the better solution, this situation has drawbacks; it needs more time and function evaluations.

In this study, in order to eliminate the drawbacks in some of previous filled functions we proposed a new filled function. This new proposal is based on putting many stationary points in lower basins, in fact, the filled function does not need to go down in the lower basin, it only needs to obtain any stationary point in the lower basin which can be used as a starting point for minimizing the objective function to get a lower minimizer. This idea helps to reduce the time and function evaluations which are very important for such methods.

This study is organized as follows: In Sect. 7.2, we give some preliminary information. In Sect. 7.3, we propose a new filled function with its properties. In Sect. 7.4, we introduce the filled function algorithm. In Sect. 7.5, we perform a numerical test and present the results obtained from the new method. Finally, Sect. 7.6 consists of conclusions.

7.2 Preliminaries

We consider a class of unconstrained global optimization problems as the following:

$$(P) \qquad \min_{x \in \mathbb{R}^n} f(x), \tag{7.1}$$

where $f(x) : R^n \longrightarrow R$ is continuously differentiable function.

We assume that the function $f(x)$ is globally convex, which means $f(x) \to +\infty$ as $\|x\| \to +\infty$. It means that there exist a closed, bounded, and box-shaped domain $\Omega = [l_b, u_b] = \{x : l_b \leq x \leq u_b,\ l_b, u_b \in \mathbb{R}^n\}$ that contains all the local minimizers of $f(x)$. Moreover, the number of different values of local minimizers of the function $f(x)$ is finite.

Additionally, basic concepts and symbols used in this study are given as follows:

k : the number of local minimizers of $f(x)$,
x_k^* : the current local minimizer of $f(x)$,

x^* : the global minimizer of $f(x)$,
B_k^* : the basin of $f(x)$ at the local minimizer x_k^*.

We indicate the following definitions:

Definition 7.1 ([34]) Let $\Omega \subset R^n$. A point $x^* \in \Omega$ is a global minimizer of objective function $f(x)$ if $f(x^*) \leq f(x)$ for all $x \in \Omega$,

Definition 7.2 ([34]) Let x_k^* is a local minimizer of the objective function $f(x)$. The set of points $B_k^* \subset \Omega$ is called a basin of $f(x)$ at the point x_k^* if any local solver starting from any point in B_k^* finds the local minimizer x_k^*.

Definition 7.3 ([34]) The auxiliary function $F(x, x_k^*)$ is called a filled function of the objective function $f(x)$ at a local minimizer x_k^* if the function $F(x, x_k^*)$ has the following properties:

- x_k^* is a local maximizer of the function $F(x, x_k^*)$,
- $F(x, x_k^*)$ has no stationary points in $A_1 = \{x \in \Omega | f(x) \geq f(x_k^*), x \neq x_k^*\}$,
- if x_k^* is not a global minimizer of the function $f(x)$, then the function $F(x, x_k^*)$ has a stationary point in the region $A_2 = \{x | f(x) < f(x_k^*), x \in \Omega\}$.

7.2.1 *Overview of the Filled Functions*

In 1987 Ge and Qin proposed a first filled function (we call it as G-function) [34] with two parameters to solve the problem (P) at an isolated local minimizer x_k^* that is defined by

$$G(x, x_k^*, r, \rho) = -(\rho^2 \ln[r + f(x)] + \|x - x_k^*\|^2), \tag{7.2}$$

and, in 1990 Ge introduced another filled function (P-function) [35] which has the following form:

$$P(x, x_k^*, r, \rho) = \frac{1}{r + f(x)} + \exp\left(-\frac{\|x - x_k^*\|^2}{\rho^2}\right), \tag{7.3}$$

where r and ρ are parameters which need to be chosen conveniently. Generally, the G-function and P-function of the objective function $f(x)$ at the current minimizer x_k^* must satisfy the Definition 3.

Many important studies are developed the FFM to solve multi-modal global optimization problems. These studies can be classified into two categories depending on the number of adjustable parameters.

7.2.2 Filled Functions with Two-Parameter

In [56], Wu et al. offer a filled function with two parameters to decrease the computational cost and overcome several disadvantages of filled functions which has the following form:

$$H_{q,r,x_k^*}(x) = q(\exp\left(-\frac{\|x - x_k^*\|^2}{q}\right)g_r(f(x)-f(x_k^*))+f_r(f(x)-f(x_k^*)), \quad (7.4)$$

where $q, r > 0$ are adjustable parameters and f_r, g_r are continuously differentiable functions.

In 2009, Zhang et al. [57] introduced a new definition for the filled function, which rectifies several drawbacks of the classic definition. A new filled function with two parameters defined by

$$P(x, x_k^*, r, a) = \varphi(r + f(x)) - a(\|x - x_k^*\|^2), \quad (7.5)$$

where $a > 0$, r are parameters and the function $\varphi(t)$ is continuously differentiable. Wei et al. [58] offer a new filled function which is not sensitive to parameters. This function has two parameters and has the following formula:

$$P(x, x_k^*) = \frac{1}{(1 + \|x - x_k^*\|^2)} g(f(x) - f(x_k^*)), \quad (7.6)$$

and

$$g(t) = \begin{cases} 0, & t \geq 0, \\ r \arctan(t^{\rho}), & t < 0, \end{cases}$$

where $r > 0$, $\rho > 1$ are parameters.

7.2.3 Filled Functions with One Parameter

According to general opinion, the existence of more than one adjustable parameter in the same filled function makes the control difficult. So, the first filled function which has only one parameter is the function Q. This function proposed in (1987) by Ge and Qin and has the following formula:

$$Q(x, a) = -(f(x) - f(x_k^*)) \exp(a\|x - x_k^*\|^2). \quad (7.7)$$

The function Q has one adjustable parameter a, if this parameter becomes large and large, quickly increasing value of exponential function negatively affects the computational results [34]. In order to tackle this drawback, H-function was introduced by Liu in [36] that is given by

$$H(x, a) = \frac{1}{\ln(1 + f(x) - f(x_k^*))} - a\|x - x_k^*\|^2. \tag{7.8}$$

The function H keeps the feature of the function Q with only one adjustable parameter but without exponential function. Shang et al. introduced a filled function with one adjustable parameter in the following:

$$F_q(x, x_k^*) = \frac{1}{(1 + \|x - x_k^*\|)}\varphi_q(f(x) - f(x_k^*) + q), \tag{7.9}$$

and

$$\varphi_q(t) = \begin{cases} \exp(-\frac{q^3}{t}), & t \neq 0, \\ 0, & t = 0, \end{cases}$$

so, q is a parameter subject to certain conditions [59]. Zhang and Xu constructed a filled function to solve non-smooth constrained global optimization problems [60]. This function constructed to overcome several drawbacks of the previous filled functions, and it has one parameter as follows:

$$P(x, x_k^*, q) = \exp(\|x - x_k^*\|)\ln(1 + q(\max\{0, f(x) - f(x_k^*) + r\}$$
$$+ \sum_{i=1}^{m} \max\{0, g_i(x)\})),$$

where $q > 0$ is the parameter, $g_i(x) > 0$ are constrained conditions, and r is prefixed constant.

In 2013, Wei and Wang proposed a new filled function for problem (P) with one adjustable parameter and it does not sensitive to this parameter [61]. The filled function has the following formula:

$$P(x, x_k^*) = -\|x - x_k^*\|^2 g(f(x) - f(x_k^*)), \tag{7.10}$$

where

$$g(t) = \begin{cases} \frac{\pi}{2}, & t \geq 0, \\ r\arctan(t^2) + \frac{\pi}{2}, & t < 0, \end{cases}$$

and r is an adjustable parameter as large as possible, used as the weight parameter.

Wang et al. constructed a filled function for both smooth and non-smooth constrained problems in 2014 [62]. The constructed filled function is defined by

$$P(x, x_k^*, q) = -\frac{1}{q}[f(x) - f(x_k^*) + \max\{0, g_i(x)\}))]^2 - \arg(1 + \|x - x_k^*\|^2)$$
$$+q[\min(0, \max(f(x) - f(x_k^*)g_i(x), i \in I))]^3.$$

The filled function at above has only one adjustable parameter which is controlled during the process. A new definition and a new filled function is given in 2016 by Yuan et al. [63]. This filled function has one parameter, given by

$$F(x, x_k^*, q) = V(\|x - x_k^*\|)W_q(f(x) - f(x_k^*)), \tag{7.11}$$

where $q > 0$ is an adjustable parameter, $V(t) : R \to R$ and $W_q(t) : R \to R$ are continuously differentiable under some properties.

7.3 A New Filled Function Method and Its Properties

We offer a new filled function at a local minimizer x_k^* with two parameters to solve the problem (P) as follows:

$$F(x, x_k^*) = \frac{1}{\alpha + \|x - x_k^*\|^2} h(f(x) - f(x_k^*)),$$

where

$$h(t) = \begin{cases} 1, & t \geq 0, \\ \sin\left(\mu t + \frac{\pi}{2}\right), & t < 0, \end{cases}$$

and $0 < \alpha \leq 1$ and $\mu > 1$ are parameters.

The new idea in this filled function is to put many stationary points in the lower basin $A_2 = \{x | f(x) < f(x_k^*), x \in \Omega\}$. In fact, the filled function does not need to go down in the lower basin, only it needs to obtain any stationary point in A_2, which can be used as an initial for minimizing objective function to obtain a lower minimizer.

The above idea has many advantages, for example, it helps to reduce the time and evaluation which are very important in cases like this. Furthermore, the parameter μ is used to increase or decrease the number of stationary points in the interval A_2, therefore we have to choose μ carefully, because if it is small there is a possibility that we may lose some of the lower minimizers at which the value of the function is close to the value at the first minimizer (see Figs. 7.1 and 7.2). The parameter $0 < \alpha \leq 1$ in the term $\frac{1}{\alpha + \|x - x_k^*\|^2}$ is used to control the hat and it is easy to modify. The following theorems references that the function $F(x, x_k^*)$ is a filled function by Definition 7.1.

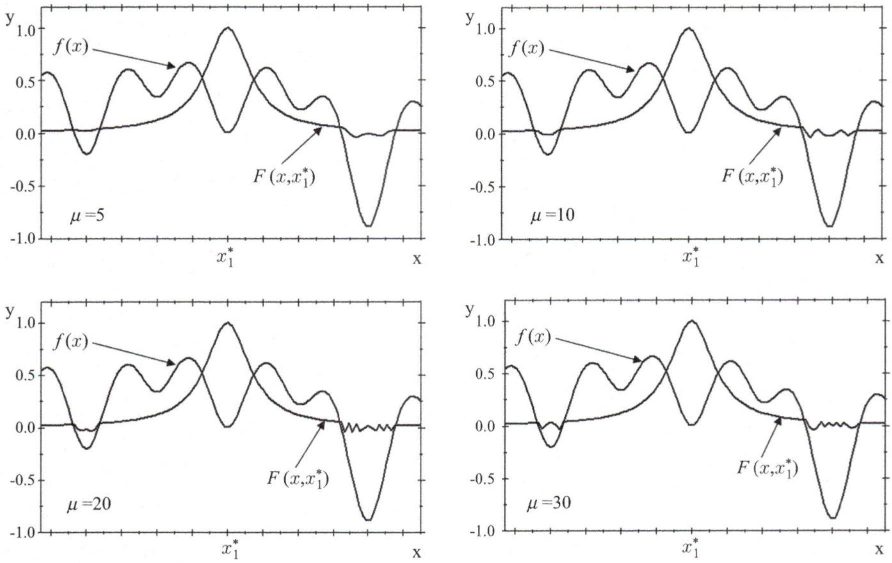

Fig. 7.1 Some different values of the parameter μ and their effect on the function $F(x, x_k^*)$

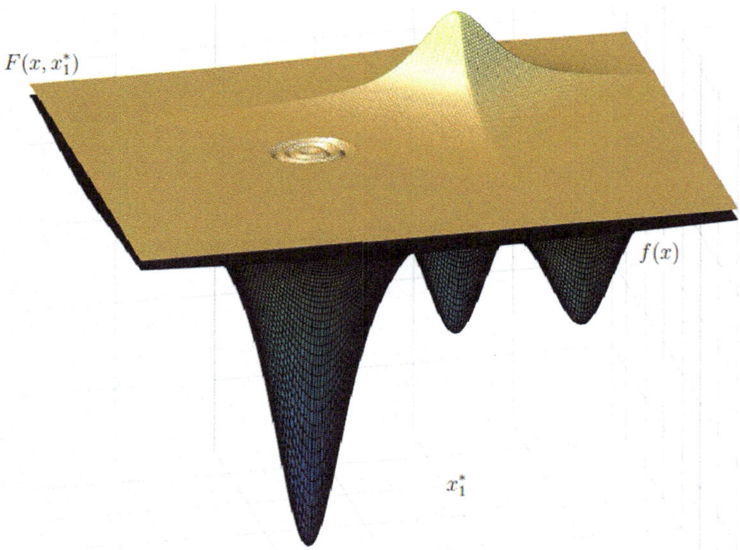

Fig. 7.2 The graph of $F(x, x_k^*)$ in two dimensions

Theorem 7.1 *Assume that x_k^* is a local minimizer of the function $f(x)$, and $F(x, x_k^*)$ is defined by the Definition 3, then x_k^* is a strict local maximizer of $F(x, x_k^*)$.*

Proof Since x_k^* is a local minimizer of $f(x)$, then there exists neighborhood $N(x_k^*, \epsilon) \subset A_1$ of x_k^* for some $\epsilon > 0$ such that $f(x) \geq f(x_k^*)$ for all $x \in N(x_k^*, \epsilon)$ and $x \neq x_k^*, 0 < \alpha \leq 1$.

$$\frac{F(x, x_k^*)}{F(x_k^*, x_k^*)} = \frac{\alpha + \|x_k^* - x_k^*\|^2}{\alpha + \|x - x_k^*\|^2} = \frac{\alpha}{\alpha + \|x - x_k^*\|^2} < 1.$$

That means x_k^* is a strict local maximizer of $F(x, x_k^*)$.

Theorem 7.2 *Assume that x_k^* is a local minimizer of $f(x)$, and x is any point in the set A_1, then x is not stationary point of $F(x, x_k^*)$ for any $0 < \alpha \leq 1$.*

Proof We have $x \in A_1$, $f(x) \geq f(x_k^*)$ and $x \neq x_k^*$. Then $F(x, x_k^*) = \frac{1}{\alpha + \|x - x_k^*\|^2}$, and $\nabla F(x, x_k^*) = -2\frac{x - x_k^*}{(\alpha + \|x - x_k^*\|^2)^2} \neq 0$, for each $0 < \alpha \leq 1$. This implies the function $F(x, x_k^*)$ has no stationary point in the set A_1.

Theorem 7.3 *Assume that $L = \min |f(x_i^*) - f(x_j^*)|$, $i, j = 1, 2, \ldots, m$, $f(x_i^*) \neq f(x_j^*)$ and x_k^* is a local minimizer of $f(x)$ but not global, then there exists a point $x' \in A_2$ such that the point x' is a stationary point of the function $F(x, x_k^*)$ when $\mu = \frac{\pi}{2L}$ for each $0 < \alpha \leq 1$.*

Proof Since the current local minimizer x_k^* is not global minimizer of $f(x)$, then there exists second minimizer $x_{k+1}^* \in A_2$ such that $f(x_{k+1}^*) < f(x_k^*)$.

For any point $y \in A_1$ we have $F(y, x_k^*) > 0$, so by the continuity of $f(x)$, and if $\mu = \frac{\pi}{2L}$ we obtain $F(x_{k+1}^*, x_k^*) < 0$. Then, by the theorem of intermediate value of continuous function, there exist a point lying between the points y and x_{k+1}^* on the part $[y, x_{k+1}^*]$, the value of the filled function at this point is equal to 0.

Assuming that z is the nearest point to x_{k+1}^* with $F(z, x_k^*) = 0$, then, we obtain the part $[z, x_{k+1}^*]$. That means $z \in \partial A_2$ and z is in the borders of the set B_{k+1}^* which is a closed region. By the continuity of the function $F(x, x_k^*)$, there exist a point $x' \in B_{k+1}^*$ such that it is a local minimizer of the function $F(x, x_k^*)$ and $F(x', x_k^*) < 0$, since the function $F(x, x_k^*)$ is continuously differentiable, we obtain

$$\nabla F(x', x_k^*) = 0.$$

7.4 The Algorithm

According to the information of the previous sections, we proposed a new filled function algorithm as follows:

1 Set $k = 1$, $\epsilon = 10^{-2}$, choose $U_\mu = 30$ an upper bound of μ and give $\mu = 5$; N the number of different directions d_i for $i = 1, 2, 3, \ldots, N$, choose an initial point $x_{int} \in \Omega$, and give $0 < \alpha \leq 1$, where n is the dimension of the problem.
2 Minimize $f(x)$ using x_{int} as a starting point to find local minimizer x_k^*.
3 Construct filled function at x_k^*

$$F(x, x_k^*) = \frac{1}{\alpha + \|x - x_k^*\|^2} h(f(x) - f(x_k^*))$$

and set $i = 1$.
4 If $i \leq N$, set $x = x_k^* + \epsilon d_i$ and go to step **(5)**; otherwise go to step **(6)**.
5 Start from x to find a minimizer x_F of $F(x, x_k^*)$, if $x_F \in \Omega$ then set $x_{int} = x_F$, $k = k + 1$ and go to step **(2)**; otherwise $i = i + 1$, go to step **(4)**.
6 If $\mu \leq U_\mu$, then $\mu = \mu + 5$ and go to step **(2)**; otherwise take x_k^* as a global minimizer of $f(x)$ and stop the algorithm.

The set of different directions d_i are as the following: let $\theta_1, \ldots, \theta_J \in [0, 2\pi]$ and $\vartheta_1, \ldots, \vartheta_J \in [-\frac{\pi}{2}, \frac{\pi}{2}]$, are uniformly distributed. If $n = 2Q$, the components of $d_i^j = (y_1^j, y_2^j, \ldots, y_{2Q}^j)$ is calculated as

$$y_{2l-1}^j = \frac{\sqrt{2}}{\sqrt{n}} \cos(\theta_j)$$

$$y_{2l}^j = \frac{\sqrt{2}}{\sqrt{n}} \sin(\theta_j),$$

for $l = 1 \sim Q$. If $n = 2Q + 1$, the components of $d_l^j = (y_1^j, y_2^j, \ldots, y_{2Q+1}^j)$ is calculated as

$$y_1^j = \frac{\sqrt{2}}{\sqrt{n}} \cos(\vartheta_j) \cos(\theta_j)$$

$$y_2^j = \frac{\sqrt{2}}{\sqrt{n}} \cos(\vartheta_j) \sin(\theta_j)$$

$$y_3^j = \frac{\sqrt{2}}{\sqrt{n}} \sin(\vartheta_j)$$

$$y_{2l}^j = \frac{\sqrt{2}}{\sqrt{n}} \cos(\theta_j)$$

$$y_{2l+1}^j = \frac{\sqrt{2}}{\sqrt{n}} \sin(\theta_j)$$

for $l = 2 \sim Q$ [64].

7.5 Numerical Results

In this section, we perform the numerical test of our algorithm on test problems which are stated as follows:

Problem 5.1–5.3 (Two-Dimensional Function)

$$\min f(x) = [1 - 2x_2 + c\sin(4\pi x_2) - x_1]^2 + [x_2 - 0.5\sin(2\pi x_1)]^2,$$

for $x_1, x_2 \in [-3, 3]$, where $c = 0.05, 0.2, 0.5$.

Problem 5.4 (Three-Hump Back Camel Function)

$$\min f(x) = 2x_1^2 - 1.05x_1^4 + \frac{1}{6}x_1^6 - x_1x_2 + x_2^2,$$

for $x_1, x_2 \in [-3, 3]$.

Problem 5.5 (Six-Hump Back Camel Function)

$$\min f(x) = 4x_1^2 - 2.1x_1^4 + \frac{1}{3}x_1^6 - x_1x_2 - 4x_2^4 + 4x_2^4,$$

for $x_1, x_2 \in [-3, 3]$.

Problem 5.6 (Treccani Function)

$$\min f(x) = x_1^4 + 4x_1^3 + 4x_1^2 + x_2^2,$$

for $x_1, x_2 \in [-3, 3]$.

Problem 5.7 (Goldstein and Price Function)

$$\min f(x) = g_1(x)g_2(x),$$

where

$$g_1(x) = 1 + (x_1 + x_2 + 1)^2(19 - 14x_1 + 3x_1^2 - 14x_2 + 6x_1x_2 + 3x_2^2),$$

and

$$g_2(x) = 30 + (2x_1 - 3x_2)^2(18 - 32x_1 + 12x_1^2 + 48x_2 - 36x_1x_2 + 27x_2^2),$$

for $x_1, x_2 \in [-3, 3]$.

Problem 5.8 (Shubert Function)

$$\min f(x) = \left\{\sum_{i=1}^{5} i\cos[(i+1)x_1 + i]\right\}\left\{\sum_{i=1}^{5} i\cos[(i+1)x_2 + i]\right\},$$

s. t. $x_1, x_2 \in [-10, 10]$.

Problem 5.9 (Rastrigin Function)

$$\min f(x) = 20 + \sum_{i=1}^{2}\left[x_i^2 - 10\cos(2\pi x_i)\right],$$

for $x_1, x_2 \in [-5.12, 5.12]$.

Problem 5.10 (Branin Function)

$$\min f(x) = \left(x_2 - \frac{5.1}{4\pi^2}x_1^2 + \frac{5}{\pi}x_1 - 6\right)^2 + 10\left(1 - \frac{1}{8\pi}\right)\cos(x_1) + 10,$$

for $x_1 \in [-5, 10],\ x_2 \in [0, 15]$.

Problems 5.11, 5.12, 5.13 (Shekel Function)

$$\min f(x) = -\sum_{i=1}^{m}\left[\sum_{j=1}^{4}(x_j - a_{i,j})^2 + b_i\right]^{-1},$$

where $m = 5, 7, 10$, b_i is an m-dimensional vector, and $a_{i,j}$ is a $4 \times m$-dimensional matrix where

$$b_i = 0.1.\left[1\ 2\ 2\ 4\ 4\ 6\ 3\ 7\ 5\ 5\right],$$

$$a_{i,j} = \begin{bmatrix} 4.0 & 1.0 & 8.0 & 6.0 & 3.0 & 2.0 & 5.0 & 8.0 & 6.0 & 7.0 \\ 4.0 & 1.0 & 8.0 & 6.0 & 7.0 & 9.0 & 3.0 & 1.0 & 2.0 & 3.6 \\ 4.0 & 1.0 & 8.0 & 6.0 & 3.0 & 2.0 & 5.0 & 8.0 & 6.0 & 7.0 \\ 4.0 & 1.0 & 8.0 & 6.0 & 7.0 & 9.0 & 3.0 & 1.0 & 2.0 & 3.6 \end{bmatrix},$$

and $x_j \in [0, 10],\ j = 1, .., 4$.

Problems 5.14–5.21 (n-Dimensional Function)

$$\min f(x) = \frac{\pi}{n}[10\sin^2 \pi x_1 + g(x) + (x_n - 1)^2],$$

where $g(x) = \sum_{i=1}^{n-1}\left[(x_i - 1)^2(1 + 10\sin^2 \pi x_{i+1})\right]$ and $x_i \in [-10, 10], i = 1, 2, \ldots, n$.

Problems 5.21–5.29 (Levy Function)

$$\min f(x) = \sin^2(\pi w_1) + \sum_{i=1}^{n-1}(w_i - 1)^2\left[1 + 10\sin^2(\pi w_i + 1)\right]$$

$$+(w_n - 1)^2\left[1 + \sin^2(2\pi w_n)\right],$$

Table 7.1 The list of test problems

Function No.	Dimension n	Function name	Optimum value	Region
5.1	2	Two-dimensional function $c = 0.05$	0	$[-3, 3]^2$
5.2	2	Two-dimensional function $c = 0.2$	0	$[-3, 3]^2$
5.3	2	Two-dimensional function $c = 0.5$	0	$[-3, 3]^2$
5.4	2	Three-hump back camel function	0	$[-3, 3]^2$
5.5	2	Six-hump back camel function	-1.0316	$[-3, 3]^2$
5.6	2	Treccani function	0	$[-3, 3]^2$
5.7	2	Goldstein and Price function	3.0000	$[-3, 3]^2$
5.8	2	Shubert function	-186.73091	$[-10, 10]^2$
5.9	2	Rastrigin function	-2.0000	$[-3, 3]^2$
5.10	2	(RC)Branin function	0.3979	$[-5, 10] \times [10, 15]$
5.11	4	$(S_{4,5})$Shekel function	-10.1532	$[0, 10]$
5.12	4	$(S_{4,7})$Shekel function	-10.4029	$[0, 10]$
5.13	4	$(S_{4,10})$Shekel function	-10.5364	$[0, 10]$
5.14–5.17	2,3,5,7	n-dimensional function	0	$[-10, 10]^2$
5.18–5.21	10,20,30,50	n-dimensional function	0	$[-10, 10]^2$
5.22–5.25	2,3,5,7	(L_5)Levy function	0	$[-10, 10]^2$
5.26–5.29	10,20,30,50	(L_7)Levy function	0	$[-10, 10]^2$

where

$$w_i = 1 + \frac{x_i - 1}{4}, \quad for \quad all \quad i = 1, \ldots, n$$

for $x_i \in [-10, 10]$, $i = 1, 2, \ldots, n$.

We rearrange the above problems and the list of test problems are presented in Table 7.1. The algorithm is implemented 10 times starting from the different points independently for each problem on a PC with Matlab R2016a. The "fminunc" function of Matlab is used as a local solver. The used symbols are the following:

- No.: the number of the problem,
- n: the dimension,
- itr-mean: the mean iteration number of the 10 runs,

Table 7.2 The numerical results of our algorithm on the list of problems

No.	$n.$	itr-mean	f-mean	f-best	f-eval	time	S-R
5.1	2	1.2000	$6.5805e-12$	$5.7244e-16$	201	0.0427	9/10
5.2	2	1.0000	$2.6536e-13$	$1.2548e-14$	315	0.0383	10/10
5.3	2	1.5000	$1.4803e-13$	$5.7321e-15$	288	0.0498	8/10
5.4	2	1.0000	$4.1081e-14$	$2.2390e-16$	306	0.0254	10/10
5.5	2	2.4000	-1.0316	-1.0316	132	0.0236	10/10
5.6	2	1.0000	$1.1315e-11$	$5.1253e-16$	240	0.0246	10/10
5.7	2	2.1000	3.0000	3.0000	414	0.0503	8/10
5.8	2	11.3000	-186.7309	-186.7309	591	0.1045	10/10
5.9	2	7.2000	$3.6948e-14$	0	246	0.0273	10/10
5.10	2	1.0000	0.3979	0.3979	243	0.0165	10/10
5.11	4	1.0000	-10.1532	-10.1532	1140	0.0614	7/10
5.12	4	3.3000	-10.4029	-10.40294	850	0.1028	9/10
5.13	4	4.1000	10.5321	-10.5321	925	0.0680	8/10
5.14	2	4.3000	$2.5282e-13$	$1.5241e-15$	150	0.0167	10/10
5.15	3	7.0000	$6.2081e-09$	$7.0745e-15$	2100	0.1873	10/10
5.16	5	9.6000	$6.5445e-09$	$2.9883e-13$	3420	0.2170	8/10
5.17	7	5.2000	$4.1802e-09$	$3.3935e-11$	5816	0.3027	8/10
5.18	10	10.1000	$4.0625e-10$	$7.1453e-12$	3850	0.1741	8/10
5.19	20	8.1000	$1.8112e-10$	$3.3503e-13$	6363	0.2280	7/10
5.20	30	5.0000	$9.3934e-11$	$2.7690e-14$	9517	0.2957	9/10
5.21	50	18.30000	$.4131e-12$	$1.8754e-15$	27846	0.8174	8/10
5.22	2	3.2000	$2.9301e-13$	$2.6872e-17$	510	0.0577	10/10
5.23	3	6.2000	$1.3768e-13$	$4.6587e-16$	1825	0.1407	9/10
5.24	5	7.4000	$1.2471e-12$	$8.6884e-14$	1422	0.0994	8/10
5.25	7	7.0000	$1.3095e-11$	$6.9033e-16$	1936	0.1151	8/10
5.26	10	8.0000	$3.6191e-12$	$5.2026e-14$	4169	0.2106	6/10
5.27	20	6.7000	$2.0352e-12$	$4.5555e-15$	7056	0.3026	6/10
5.28	30	14.7000	$.8459e-12$	$7.1327e-16$	13391	0.6136	10/10
5.29	50	10.4000	$9.9071e-12$	$9.8943e-15$	48450	1.1783	9/10

- f-eval: the total number of function evaluations,
- time: the mean of the aggregate running time in 10 runs (second),
- f-mean: the mean of the function values in 10 runs,
- f-best: the best function value in 10 runs,
- S-R: the success rate among 10 implementation with different starting points.

The results of the algorithm is presented in Table 7.2. This table consists of seven column; (No) number of the problem, (n) dimension of the problem, (itr-mean) mean value of total iteration number, (f-mean) mean value of the function values, (f-best) the best value of the function values, (f-eval) mean value of the function evaluations and (S-R) success rates of ten trails uniform initial points. As shown in this table our algorithm is tested on 29 problems with different dimensions up to

Table 7.3 The comparison of our algorithm with algorithm in [65]

		Our method			The method in [65]		
No.	n.	f-best	f-eval	S-R	f-best	f-eval	S-R
5.1	2	$5.7244e-16$	201	$9/10$	$2.66630e-15$	214	$8/10$
5.2	2	$1.2548e-14$	315	$10/10$	$3.4336e-16$	290.6250	$8/10$
5.3	2	$5.7321e-15$	288	$8/10$	$4.7243e-16$	414.2857	$8/10$
5.4	2	$2.2390e-16$	306	$10/10$	$2.8802e-16$	411	$10/10$
5.5	2	-1.0316	132	$10/10$	-1.0316	234	$10/10$
5.6	2	$5.1253e-16$	240	$10/10$	$1.6477e-15$	216.5000	$10/10$
5.7	2	3.0000	414	$8/10$	3.0000	487.8889	$9/10$
5.8	2	-186.7309	591	$10/10$	-186.7309	813.5000	$10/10$
5.9	2	0	246	$10/10$	-2.0000	501	$10/10$
5.10	2	0.3979	243	$10/10$	0.3979	222.3000	$10/10$
5.11	4	-10.1532	1140	$7/10$	-10.1532	1001	$9/10$
5.12	4	-10.4029	850	$9/10$	-10.4029	1365.1000	$8/10$
5.13	4	-10.5321	925	$8/10$	-10.5321	1412	$7/10$
5.14	2	$1.5241e-15$	150	$10/10$	$9.4192e-15$	743	$8/10$
5.15	3	$7.0745e-15$	2100	$10/10$	$5.6998e-15$	3027	$10/10$
5.16	5	$2.9883e-13$	3420	$8/10$	$3.7007e-15$	4999	$10/10$
5.17	7	$3.3935e-11$	5816	$8/10$	$1.3790e-14$	8171	$8/10$
5.18	10	$7.1453e-12$	3850	$8/10$	$3.0992e-14$	8895	$9/10$
5.19	20	$3.3503e-13$	6363	$7/10$	$3.0016e-13$	18242	$7/10$
5.20	30	$2.7690e-14$	9517	$9/10$	$1.7361e-12$	43232	$6/10$
5.21	50	$1.8754e-15$	27846	$8/10$	$9.8531e-13$	83243	$6/10$

50 dimensions, each problem is tested on ten different initial points. Our algorithm reaches $10/10$ success rate for almost 40% of the all test problems. At least $6/10$ success rate is obtained considering all of the test problems as stated in column (S-R). By using our algorithm %94 success rate is obtained considering the total number of trials. Moreover, the f-mean and f-best values in Table 7.2 are very close to original function values which are given in Table 7.1.

The comparison of our algorithm with the method in [65] are summarized in Table 7.3.

In general, it can be seen from Table 7.3 the results of the algorithm presented in this paper obtain an advantage in several places especially in the columns dedicated to function evaluations and success rates compared to the results of the algorithm in [65]. Both of the methods (our method and the method in [65]) are sufficiently successful in terms of "f-best" values. Our method complete the global optimization process with lower function evaluation values than the method in [65] for the 85% of the all test problems. By using our algorithm, 88% successful trial is obtained considering the total number of the trial used in comparison and at least $7/10$ success rate is obtained for each of the test problems used in comparison. 86% successful trial is obtained considering the total number of trials used in comparison and at least $7/10$ success rate is obtained for each of the test problems in comparison, by using the algorithm in [65]. Moreover, our method is more effective than the method in [65] in terms of "f-eval" and "S-R" on the 10 and more than 10 dimensional test problems. Thus, the introduced algorithm in this paper is more efficient than the algorithm introduced in [65].

7.6 Conclusions

In this chapter, a new filled function for unconstrained global optimization is presented and the useful properties are introduced. The proposed filled function contains two parameters which can be easily adjusted in the minimization process. The corresponding filled function algorithm is constructed. Furthermore, it has been performed on numerical experiment in order to demonstrate the effectiveness of the presented algorithm. It can be seen from the computational results that the present method is promising.

This new algorithm is an effective approach to solve multi-modal global optimization problems. In the minimizing process, our methods save time in finding the lower minimizer and it is guaranteed that upper minimizers are not taken into account in all of the minimization process independently from the parameters. These two important properties make our method advantageous among the other methods.

For future work, the applications of global optimization algorithms can be applied to many real-life problems such as data mining, chemical process, aerospace industries, etc.

References

1. Savku, E., Weber, G-W.: A stochastic maximum principle for a Markov regime-switching jump-diffusion model with delay and an application to finance. J. Optim. Theory Appl. **179**, 696–721 (2018)
2. Yazici, C., Yerlikaya-Ozkurt, F., Batmaz, I.: A computational approach to nonparametric regression: bootstrapping CMARS method. Mach. Learn. **101**, 211–230 (2015)
3. Kara, G., Ozmen A., Weber, G-W.: Stability advances in robust portfolio optimization under parallelepiped uncertainty. Cent. Eur. J. Oper. Res. **27**, 241–261 (2019)
4. Onak, O.N., Serinagaoglu-Dogrusoz, Y., Weber, G.-W.: Effects of a priori parameter selection in minimum relative entropy method on inverse electrocardiography problem. Inverse Prob. Sci. Eng. **26**(6), 877–897 (2018)
5. Resener, M., Haffner, S., Pereira, L.A., Pardalos, P.A.: Optimization techniques applied to planning of electric power distribution systems: a bibliographic survey. Energy Syst. **9** 473–509 (2018)
6. Addis, B., Cassioli, A., Locatelli, M., Schoen, F.: A global optimization method for the design of space trajectories. Comput. Optim. Appl. **48**, 635–652 (2011)
7. Cassioli, A., Di Lorenzo, D., Locatelli, M., Schoen, F., Sciandrone, M.: Machine learning for global optimization. Comput. Optim. Appl. **51**, 279–303 (2012)
8. Chen, T.Y., Huang, J.H.: Application of data mining in a global optimization. Adv. Eng. Softw. **66**, 24–33 (2013)

9. Leary, R.H.: Global optimization on funneling landscapes. J. Glob. Optim. **18**, 367–383 (2000)
10. Locatelli, M., Schoen, F.: Global optimization based on local searches. Ann. Oper. Res. **240**, 251–270 (2016)
11. Land, A.H., Doig, A.G.: An automatic method of solving discrete programming problems. Econometrica **28**, 497–520 (1960)
12. Jones, D.R. , Perttunen, C.D. , Stuckman, B.E.: Lipschitzian optimization without the Lipschitz constant. J. Optimiz. Theory Appl. **79**, 157–181 (1993)
13. Zhigljavski, A., Zilinskas, J.: Stochastic Global Optimization. Springer, Berlin (2008)
14. Schaffler, S.: Global Optimization: A Stochastic Approach. Springer, New York (2012)
15. Storti, G.L., Paschero, M., Rizzi, A., Mascioli, F.M.: Comparison between time-constrained and time-unconstrained optimization for power losses minimization in smart grids using genetic algorithms. Neurocomputing **170**, 353–367 (2015)
16. Suman, B., Kumar, P.: A survey of simulated annealing as a tool for single and multiobjective optimization. J. Oper. Res. Soc. **57**, 1143–1160 (2006)
17. Ekren, O., Ekren, B.Y.: Size optimization of a PV/wind hybrid energy conversion system with battery storage using simulated annealing. Appl. Energ. **87**, 592–598 (2010)
18. Samora, I., Franca, M.J., Schleiss, A.J., Ramos, H.M.: Simulated annealing in optimization of energy production in a water supply network. Water Resour. Manag. **30**, 1533–1547 (2016)
19. Poli, R., Kennedy, J., Blackwell, T.: Particle swarm optimization. Swarm Intell-US. **1**, 33–57 (2007)
20. Kennedy, J.: Particle swarm optimization. In: Encyclopedia of Machine Learning, pp. 760–766. Springer, Boston (2011)
21. Karaboga, D., Basturk, B.: A powerful and efficient algorithm for numerical function optimization: artificial bee colony (ABC) algorithm. J. Global Optim. **39**, 459–471 (2007)
22. Akay, B., Karaboga, D.: A modified artificial bee colony algorithm for real-parameter optimization. Inform. Sciences. **192**, 120–142 (2012)
23. Niknam, T., Amiri, B., Olamaei, J., Arefi, A.: An efficient hybrid evolutionary optimization algorithm based on PSO and SA for clustering. J. Zhejiang Univ-Sc. A. **10**, 512–519 (2009)
24. Mahi, M., Baykan, O.K., Kodaz, H.: A new hybrid method based on particle swarm optimization, ant colony optimization and 3-opt algorithms for traveling salesman problem. Appl. Soft Comput. **30**, 484–490 (2015)
25. Zheng, Y.J., Xu, X.L., Ling, H.F., Chen, S.Y.: A hybrid fireworks optimization method with differential evolution operators. Neurocomputing **148**, 75–82 (2015)
26. Garg, H.: A hybrid PSO-GA algorithm for constrained optimization problems. Appl. Math. Comput. **274**, 292–305 (2016)
27. Liu, J., Zhang, S., Wu, C., Liang, J., Wang, X., Teo, KL.: A hybrid approach to constrained global optimization. Appl. Soft. Comput. **47**, 281–294 (2016)
28. Sergeyev, Y.D., Kvasov, D.E.: A deterministic global optimization using diagonal auxiliary functions. Commun. Nonlinear Sci. Numer. Simul. **21**, 99–111 (2015)
29. Lera, D., Sergeyev, Y.D.: Deterministic global optimization using space filling curves and multiple estimates of Lipschitz and Hölder constants. Commun. Nonlinear Sci. Numer. Simul. **23**, 328–342 (2015)
30. Ziadi, R., Bencherif-Madani, A., Ellaia, R.: Continuous global optimization through the generation of parametric curves. Appl. Math. Comput. **282**, 65–83 (2016)
31. Basso, P.: Iterative methods for the localization of the global maximum. SIAM J. Numer. Anal. **19**, 781–792 (1982)
32. Mladineo, R.H.: An algorithm for finding the global maximum of a multimodal, multivariate function. Math. Program. **34**, 188–200 (1986)
33. Levy, A.V., Montalvo, A.: The tunneling algorithm for the global minimization of functions. SIAM J. Sci. Stat. Comput. **6**, 15–29 (1985)
34. Ge, R.P., Qin, Y.F.: A class of filled functions for finding global minimizers of a function of several variables. J. Optimiz. Theory. App. **54**, 241–252 (1987)
35. Ge, R.P.: A filled function method for finding a global minimizer of a function of several variables. Math. Program. **46**, 191–204 (1990)

36. Liu, X.: Finding global minima with a computable filled function. J. Global. Optim. **19**, 151–161 (2001)
37. Wu, Z.Y., Li, D., Zhang, L.S.: Global descent methods for unconstrained global optimization. J. Global. Optim. **50**, 379–396 (2011)
38. Cetin, B.C., Barhen, J., Burdick, J.W.: Terminal repeller unconstrained subenergy tunneling (TRUST) for fast global optimization. J. Optim. Appl. **77**, 97–126 (1993)
39. Groenen, P.J., Heiser, W.J.: The tunneling method for global optimization in multidimensional scaling. Psychometrika **61**, 529–550 (1996)
40. Chowdhury, P.R., Singh, Y.P., Chansarkar, R.A.: Hybridization of gradient descent algorithms with dynamic tunneling methods for global optimization. IEEE T. Syst. Man Cy. A. **30**, 384–390 (2000)
41. Xu, Y.T., Zhang, Y., Wang, S.G.: A modified tunneling function method for non-smooth global optimization and its application in artificial neural network. Appl. Math. Model. **39**, 6438–6450 (2015)
42. Xu, Z., Huang, H.X., Pardalos, P.M., Xu, C.X.: Filled functions for unconstrained global optimization. J. Global. Optim. **20**, 49–65 (2001)
43. Wu, Z.Y., Zhang, L.S., Teo, K.L., Bai, F.S.: New modified function method for global optimization. J. Optim. Theory App. **125**, 181–203 (2005)
44. Wu, Z.Y., Bai, F.S., Lee, H.W., Yang, Y.J.: A filled function method for constrained global optimization. J. Global Optim. **39**, 495–507 (2007)
45. Zhang, Y., Zhang, L., Xu, Y.: New filled functions for nonsmooth global optimization. Appl. Math. Model. **33**, 3114–3129 (2009)
46. Sahiner, A., Gokkaya, H., Yigit, T.: A new filled function for nonsmooth global optimization. In: AIP Conference Proceedings, pp. 972–974. AIP (2012)
47. Wang, W., Zhang, X., Li, M.: A filled function method dominated by filter for nonlinearly global optimization. J. Appl. Math. (2015). doi: https://doi.org/10.1155/2015/245427
48. Yuan, L.Y., Wan, Z.P., Tang, Q.H., Zheng, Y.: A class of parameter-free filled functions for box-constrained system of nonlinear equations. Acta Math. Appl. Sin-E. **32**, 355–64 (2016)
49. Wei, F., Wang, Y., Lin, H.: A new filled function method with two parameters for global optimization. J. Optim. Theory. App. **163**, 510–527 (2014)
50. Lin, H., Gao, Y., Wang, Y.: A continuously differentiable filled function method for global optimization. Numer. Algorithms **66**, 511–523 (2014)
51. Yilmaz, N., Sahiner, A.: New global optimization method for non-smooth unconstrained continuous optimization. In: AIP Conference Proceedings, pp. 250002. AIP (2017)
52. Sahiner, A., Yilmaz, N., Kapusuz, G.: A descent global optimization method based on smoothing techniques via Bezier curves. Carpathian J. Math. **33**, 373–380 (2017)
53. Lin, H., Wang, Y., Gao, Y., Wang, X.: A filled function method for global optimization with inequality constraints. Comput. Appl. Math. **37**, 1524–1536 (2018)
54. Liu, H., Wang, Y., Guan, S., Liu, X.: A new filled function method for unconstrained global optimization. Int. J. Comput. Math. **94**, 2283–2296 (2017)
55. Sahiner, A., Ibrahem, S.A.: A new global optimization technique by auxiliary function method in a directional search. Optim. Lett. (2018). doi: https://doi.org/10.1007/s11590-018-1315-1
56. Wu, Z.Y., Lee, H.J., Zhang, L.S., Yang, X.M.: A novel filled function method and quasi-filled function method for global optimization. Comput. Optim. Appl. **34**, 249–272 (2005)
57. Zhang, Y., Zhang, L., Xu, Y.: New filled functions for nonsmooth global optimization. Appl. Math. Model. **33**, 3114–3129 (2009)
58. Wei, F., Wang, Y., Lin, H.: A new filled function method with two parameters for global optimization. J. Optim. Theory App. **163**, 510–527 (2014)
59. Shang, Y.L., Pu, D.G., Jiang, A.P.: Finding global minimizer with one-parameter filled function on unconstrained global optimization. Appl. Math. Comput. **191**, 176–182 (2007)
60. Zhang, Y., Xu, Y.T.: A one-parameter filled function method applied to nonsmooth constrained global optimization. Comput. Math. Appl. **58**, 1230–1238 (2009)
61. Wei, F., Wang, Y.: A new filled function method with one parameter for global optimization. Math. Probl. Eng. (2013). doi: https://doi.org/10.1155/2013/532325

62. Wang, W.X., Shang, Y.L., Zhang, Y.: Global minimization of nonsmooth constrained global optimization with filled function. Math. Probl. Eng. (2014). doi: https://doi.org/10.1155/2014/563860
63. Yuan, L., Wan, Z., Tang, Q.: A criterion for an approximation global optimal solution based on the filled functions. J. Ind. Manag. Optim. **12**, 375–387 (2016)
64. Wang, Y., Fan, L.: A smoothing evolutionary algorithm with circle search for global optimization. In: 4th IEEE International Conference, pp. 412–418 (2010)
65. Sahiner, A., Yilmaz, N., Kapusuz, G.: A novel modeling and smoothing technique in global optimization. J. Ind. Manag. Optim. (2018). doi: https://doi.org/10.3934/jimo.2018035

Chapter 8
A New Approach for the Solution of the Generalized Abel Integral Equation

Tahir Cosgun, Murat Sari, and Hande Uslu

8.1 Introduction

Singular integral equations are examples of the topics included in the applied fields before the theory. Abel tried to answer one of the well-known problems of mechanics in the opposite direction [1, 2]. Instead of computing the required time for a particle that slides on a given trajectory, Abel posed the question: For given two points A and B in the plane, can we find the trajectories, combining the points A and B, that requires the same amount of time? This problem is called as tautochrone problem and has a deserved reputation in both mathematics and mechanics. The name tautochrone combining two words tauto (same) and chrono (time) comes from Latin. The mathematical formulation of this problem is as follows:

Suppose that a particle with mass m slides downward through a frictionless surface, i.e., the only force acting on the mass m is the gravitational force. See Fig. 8.1 for a pictorial explanation. Assume the rate of change of the arc with respect to the height is

$$\frac{ds}{dy} = f(y). \tag{8.1}$$

Considering the conservation of energy, the change in the potential must be equal to the change in the kinetic energy

T. Cosgun (✉)
Amasya University, Amasya, Turkey

Yildiz Technical University, Istanbul, Turkey
e-mail: tahir.coskun@amasya.edu.tr

M. Sari · H. Uslu
Yildiz Technical University, Istanbul, Turkey
e-mail: sarim@yildiz.edu.tr; usluh@yildiz.edu.tr

© Springer Nature Switzerland AG 2020

J. A. T. Machado et al. (eds.), *Numerical Solutions of Realistic Nonlinear Phenomena*, Nonlinear Systems and Complexity 31, https://doi.org/10.1007/978-3-030-37141-8_8

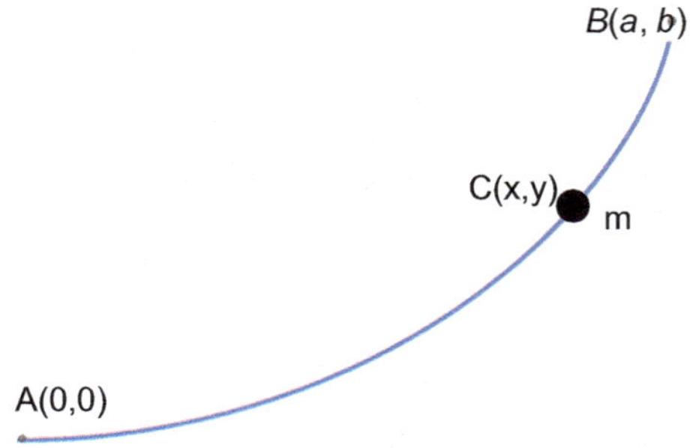

Fig. 8.1 The particle with mass m slides through a frictionless path from point B to point A under the influence of gravity

$$mgb - mgy = \frac{1}{2}mv^2. \tag{8.2}$$

Canceling m's and solving the last equation for v, we can obtain

$$v = \sqrt{2g(b-y)}. \tag{8.3}$$

Now, using the chain rule we can also observe that

$$\frac{ds}{dy} = \frac{ds}{dt}/\frac{dy}{dt} = \frac{vdt}{dy}. \tag{8.4}$$

Hence, using Eqs. (8.1) and (8.3), and rearranging the terms one can deduce the following equalities

$$dt = \frac{f(y)}{v}dy \tag{8.5}$$

$$= \frac{f(y)}{\sqrt{2g(b-y)}}dy. \tag{8.6}$$

Integrating both sides of Eq. (8.6) from 0 to b, one of the well-known integral equations can be obtained as

$$T(b) = \int_0^b \frac{f(y)}{\sqrt{2g(b-y)}}dy. \tag{8.7}$$

Notice that, in Eq. (8.7), $T(b)$ is a given time, and the only unknown is the trajectory $f(y)$ under the integral sign. With an appropriate change of variables Eq. (8.7) can be transformed into a more familiar form

$$f(x) = \int_0^x \frac{u(t)}{\sqrt{x-t}}dt. \tag{8.8}$$

To solve this equation, a more generalized version of this equation was solved by Abel:

$$f(x) = \int_0^x \frac{u(t)}{(x-t)^\alpha} dt, \tag{8.9}$$

where $\alpha \in (0, 1)$. Throughout this study, we will deal with this form of the problem which is known as the generalized Abel integral equation. In the literature, this equation is sometimes called the generalized Volterra integral equation of the first kind [3]. It is also strictly connected with fractional integrals.

In the literature, there are various techniques to solve Eq. (8.9). For example, a spectral iterative method [4], an operational matrix method [5], an operational method via Jacobi polynomials [6], a numerical solution via product trapezoidal method [7], a solution by using the normalized Bernstein polynomials [8], and a method based on regularization [9] have been proposed to solve Eq. (8.9). Being not limited to the mentioned methods, there are also other techniques proposed in the literature to solve the problem [10–15].

The main reason for focusing so much attention of researchers on this topic is application facilities. In fact, both Volterra and Fredholm integral equations of the first kinds are firmly connected to inverse problems. In the solution procedures of inverse problems scientifically important models are tried to be obtained through the physical observation of scattered data. It is hard or sometimes even impossible to determine this kind of models directly. In this respect, X-ray radiography [16], plasma spectroscopy [17], radar ranging [18], atomic scattering [19], electron emission [20], seismology [21], and microscopy [22] could be given as some application areas where the Abel integral equations are utilized.

8.2 Derivations and Main Results

In the present solution procedure, we want to replace the singular kernel of Eq. (8.9)

$$K(x, t) := \frac{1}{(x-t)^\alpha}, \qquad 0 < \alpha < 1 \tag{8.10}$$

with a series expansion via the generalized binomial theorem. Therefore, first of all, we want to remind the generalized binomial theorem.

Theorem 8.1 *For any $x, y \in R$*

$$(x+y)^r = \sum_{k=0}^{\infty} \binom{r}{k} x^k y^{r-k}, \tag{8.11}$$

where r is an arbitrary real number, converges absolutely for $|x/y| < 1$.

Proof Proof can be found in [23, 24].

As a remark, let us emphasize the definition of the generalized binomial coefficients:

$$\binom{r}{k} := \frac{r(r-1)\dots(r-k+1)}{k!}. \tag{8.12}$$

Now, being equipped with the generalized binomial theorem we can make the following observations:

$$\frac{1}{(x-t)^{\alpha}} = \sum_{k=0}^{\infty} \binom{-\alpha}{k} (-t)^k (x)^{-\alpha-k}. \tag{8.13}$$

Note that this series converges absolutely for $|t/x| < 1$. After the approximation to the kernel, we need to assume the analyticity of the source function $f(x)$, i.e., it possesses a Taylor series expansion:

$$f(x) = \sum_{k=0}^{\infty} f_k x^k, \tag{8.14}$$

and we suppose the unknown function $u(x)$ has the form

$$u(x) = x^{\alpha-1} \sum_{m=0}^{\infty} u_m x^m, \tag{8.15}$$

where f_k's and u_m's are just the coefficients in their series expansions. Hence, the problem reduces to find u_m's. Now, plugging Eqs. (8.13)–(8.15) into Eq. (8.9) we can obtain

$$\sum_{k=0}^{\infty} f_k x^k = \int_0^x \left(\sum_{k=0}^{\infty} \binom{-\alpha}{k} (-t)^k (x)^{-\alpha-k} \right) \left(\sum_{m=0}^{\infty} u_m t^{m+\alpha-1} \right) dt. \tag{8.16}$$

If we take two generic terms from the series in the right-hand side of Eq. (8.16) and integrate their multiplication over the interval $(0, x)$, we can obtain the following:

$$\begin{aligned}
\int_0^x \binom{-\alpha}{k} (-t)^k (x)^{-\alpha-k} u_m t^{m+\alpha-1} dt &= \binom{-\alpha}{k} (-1)^k (x)^{-\alpha-k} u_m \int_0^x t^{k+m+\alpha-1} dt \\
&= \binom{-\alpha}{k} (-1)^k (x)^{-\alpha-k} u_m \frac{x^{k+m+\alpha}}{k+m+\alpha} \\
&= \frac{\binom{-\alpha}{k} (-1)^k u_m x^m}{k+m+\alpha}.
\end{aligned}$$

Notice that the remaining term includes only x^m. Therefore, instead of multiplying just two terms from two different series, since the power of x is independent of k's in the above calculation, we can multiply the general term $u_m t^{m+\alpha-1}$ with the series expansion of the kernel and then integrating we can obtain

$$\sum_{m=0}^{\infty} f_m x^m = \sum_{m=0}^{\infty} u_m \left(\sum_{k=0}^{\infty} \binom{-\alpha}{k} \frac{(-1)^k}{k+m+\alpha} \right) x^m$$
$$= \sum_{m=0}^{\infty} \frac{u_m \Gamma(1-\alpha)\Gamma(m+\alpha+1)}{(\alpha+m)\Gamma(m+1)} x^m.$$

Here, Γ represents the usual gamma function. Thus, by considering the same powers of x in the last equation we can find the coefficients in the series expansion of $u(x)$ as

$$u_m = \frac{f_m(\alpha+m)\Gamma(m+1)}{\Gamma(1-\alpha)\Gamma(m+\alpha+1)}. \tag{8.17}$$

As a result, we can obtain the desired solution of the generalized Abel integral equation as

$$u(x) = \frac{1}{\Gamma(1-\alpha)} \sum_{m=0}^{\infty} \frac{f_m(\alpha+m)\Gamma(m+1)}{\Gamma(m+\alpha+1)} x^{m+\alpha-1}. \tag{8.18}$$

8.3 Illustrative Examples

Example 8.1 Consider the generalized Abel integral equation

$$x^n = \int_0^x \frac{1}{(x-t)^\alpha} u(t)dt, \tag{8.19}$$

where n is a given integer.

Solution 8.1 The solution of the given equation can be obtained analytically as

$$u(x) = \frac{(\alpha+n)\Gamma(n+1)}{\Gamma(1-\alpha)\Gamma(n+\alpha+1)} x^{n+\alpha-1}.$$

□

From Example 8.1, we can deduce that whenever the given source function $f(x)$ is a polynomial of finite degree, the unknown solution $u(x)$ can be determined analytically.

Example 8.2 Consider the generalized Abel integral equation

$$\sin(x) = \int_0^x \frac{1}{(x-t)^\alpha} u(t)dt. \tag{8.20}$$

Solution 8.2 The solution can be obtained by

$$u(x) = \sum_{n=0}^{\infty} \frac{(-1)^n(\alpha+2n+1)\Gamma(2n+2)}{(2n+1)!\Gamma(1-\alpha)\Gamma(2n+2+\alpha)} x^{2n+\alpha}.$$

□

Example 8.3 Consider the generalized Abel integral equation

$$e^x = \int_0^x \frac{1}{(x-t)^\alpha} u(t)dt. \tag{8.21}$$

Solution 8.3 Then, the solution can be obtained by

$$u(x) = \sum_{n=0}^{\infty} \frac{(\alpha+n)\Gamma(n+1)}{n!\Gamma(1-\alpha)\Gamma(n+\alpha+1)} x^{n+\alpha-1}.$$

□

In Examples 8.2 and 8.3, the given source functions are obviously analytical functions and can be expanded into Taylor series. This was one of the assumptions in Sect. 8.2. The method presented in this study has a major advantage when $f(x)$ is an analytical function. On the other hand, this kind of assumptions may be considered as a drawback for the proposed method. For example, if n is not a natural number in the first example, of course, it is possible because of Stone–Weierstrass theorem, but it will be a cumbersome task to approximate $f(x)$ with a polynomial.

8.4 Conclusions

To reach a solution to an equation in different ways or via different solution procedures is an important task for scientists. Actually, it is not a task only, but also a pleasure. In this study, we have obtained the solution of the generalized Abel integral equation via a numerical solution process, and we have obtained the same results with Abel. We hope that this study will stimulate enthusiasts to use the generalized Binomial theorem when they are solving singular integral equations or treating fractional integrals numerically.

References

1. Abel, N.H.: Oplósning af et par opgaver ved hjelp af bestemte integraler. Magazin for Naturvidenskaberne **1**(2), Christiania (1823)
2. Abel, N.H.: Auflösung einer mechanischen ausgabe. J. für die Reine und Angewandte Mathematik. **1**, 153–157 (1826)
3. Trichomi, F.G.: Integral Equations. Dover, New York (1982)
4. Shoja, A., Vahidi, A.R., Babolian, E.: A spectral iterative method for solving nonlinear singular Volterra integral equations. Appl. Numer. Math. **112**, 79–90 (2017)
5. Dixit, S., Pandey, R.K., Kumar, S., Singh, O.P.: Solution of the generalized Abel integral equation by using almost Bernstein operational matrix. Am. J. Comp. Math. **1**, 226–234 (2011)
6. Sadri, K., Amini, A., Cheng, C.: A new operational method to solve Abel's and generalized Abel's integral equations. Appl. Math. Comput. **317**, 49–67 (2018)
7. Atkinson, K.E.: The numerical solution of an Abel integral equation by a product trapezoidal method. SIAM J. Numer. Anal. **11**(1), 97–101 (1974)
8. Singh, V.K., Pandey, R.K., Singh, O.P.: New stable solutions of singular integral equations of Abel type by using normalized Bernstein polynomials. Appl. Math. Sci. **3**(5), 241–255 (2009)
9. Shou, H.: Application of the regularization method to the numerical solution of Abel's integral equation. J. Comput. Math. **3**(1), 27–34 (1985)
10. Chakrabarti, A., George, A.J.: Diagonalizable generalized Abel integral operators. SIAM J. Appl. Math. **57**(2), 568–575 (1997)
11. Pandey, R.K., Sharma, S., Kumar, K.: Collocation method for generalized Abel's integral equations. J. Comput. Appl. Math. **302**, 118–128 (2016)
12. Murio, D.A., Hinestroza, G.D., Mejia, C.E.: New stable numerical inversion of Abel's integral equation. Comput. Math. Appl. **23**(11), 3–11 (1992)
13. Huang, L., Huang, Y., Li, X.F.: Approximate solution of Abel integral equations. Comput. Math. Appl. **56**, 1748–1757 (2008)
14. Vanani, S.K., Soleymani, F.: Tau approximate solution of weakly singular Volterra integral equations. Math. Comput. Model. **57**, 494–502 (2013)
15. Chakrabarti, A.: Solution of the generalized Abel integral equation. J. Integral Equ. Appl. **20**(1) (2008)
16. Deutsch, M., Notea, A., Pal, D.: Inversion of Abel's integral equation and its application to NDT by X-ray radiography, NDT. International **23**(1) (1990)
17. Fleurier, C., Chapelle, J.: Inversion of Abel's integral equation application to plasma spectroscopy. Comput. Phys. Commun. **7**, 200–206 (1974)
18. Hellsten, H., Andersson, L.E.: An inverse method for the processing of synthetic aperture radar data. Inverse Problems **3**, 111–124 (1987)
19. Buck, U.: Inversion of molecular scattering data. Rev. Mod. Phys. **46**, 369–389 (1974)
20. Kosarev, E.L.: Applications of integral equations of the first kind in experiment physics. Comput. Phys. Commun. **20**, 69–75 (1980)
21. Macelwane, J.B.: Evidence on the Interior of the Earth Derived from Seismic Sources, pp. 227–304. Internal Constitution of the Earth B. Gutenberg, Dover, New York (1951)
22. Jakeman, A.J., Anderssen, R.S.: Abel type integral equations in stereology I. General discussion. J. Microsc. **105**, 121–133 (1975)
23. Bromwich, T.J.I'A: An Introduction to the Theory of Infinite Series. Macmillan, London (1908)
24. Graham, R.L., Knuth, D.E., Patashnik, O.: Concrete Mathematics: A Foundation for Computer Science. Addison-Wesley (1994)

Chapter 9
NPSOG: A New Hybrid Method for Unconstrained Differentiable Optimization

Halima Lakhbab

9.1 Introduction

Optimization is essentially the art, science, and mathematics of choosing the best among a given set of finite or infinite alternatives.

To use optimization, we must first identify some objective, a quantitative measure of the performance of the system under study. The objective depends on certain characteristics of the system, called variables or unknowns. We must also identify constraints for the given problem. This step is known as *modeling*.

Once the model has been formulated, an optimization algorithm can be used to find its solution. The effectiveness of the results of the application of any optimization algorithm is largely a function of the degree to which the model represents the system studied. The choice of an optimization algorithm is also an important task; it may determine whether the problem is solved rapidly or slowly and, indeed, whether the solution is found at all.

So, the goal of optimization is, for a given problem formalized as an objective function $f : D \longrightarrow E$, to find the element of D which gives the best solution in E.

The space D can be called the search space, the problem space, or the function's domain. The element $x \in D$ may be named a vector, a solution, a variable, or possibly a point in D.

In this work, we will consider nonlinear continuous unconstrained optimization problems, in which $D = \mathbb{R}^n$, $E = \mathbb{R}$, and f is a nonlinear differentiable function.

H. Lakhbab (✉)
LIMSAD Laboratory, Department of Mathematics and Computer Science, Faculty of Sciences Aïn Chock - Casablanca, Hassan II University, Casablanca, Morocco

© Springer Nature Switzerland AG 2020

J. A. T. Machado et al. (eds.), *Numerical Solutions of Realistic Nonlinear Phenomena*, Nonlinear Systems and Complexity 31, https://doi.org/10.1007/978-3-030-37141-8_9

The unconstrained minimization problem

$$\min_{x \in \mathbb{R}^n} f(x)$$

has different iterative methods to solve it: If x_k denotes the current iterate, and if it is not a good estimator of the solution x_*, a better one, $x_{k+1} = x_k - \alpha_k g_k$ is required. Here g_k is the gradient vector of f at x_k and the scalar α_k is the step length.

Our main contribution in this chapter is to develop a novel hybrid approach based on a nonmonotone spectral gradient method and particle swarm optimization.

Nonmonotone spectral gradient (NSG) techniques are considered for unconstrained optimization of differentiable functions. They combine a nonmonotone steplength strategy that is based on the Grippo–Lampariello–Lucidi nonmonotone line search [14] with the spectral gradient choice of steplength [3]. This method requires few storage locations and inexpensive computations. Furthermore the nonmonotone line search assures the global convergence.

The fact that this method has good local search characteristics motivates us to combine it with particle swarm optimization to approach the global minimum.

Traditionally a hybrid PSO carries out first a certain number of iteration and then an iterative method (Gradient type methods or another metaheuristic) is applied to refine the approximations. On the contrary, in our new hybrid approach, in every iteration of PSO, and under specific condition given by the notion of loudness parameter, we perform an exploitation step by NSG method.

The remainder of this chapter is organized as follows: In Sect. 9.2 we introduce new gradient-based method, considered for the minimization of differentiable functions. This method combines two recently developed ingredients in optimization: the nonmonotone line search schemes and a spectral steplength. At first, we present the "Barzilai–Borwein method" [3], known also by spectral gradient method, this method consists essentially of a gradient descent method, where the choice of the step size along the antigradient direction is potentially derived from a two-point approximation to the secant equation underlying the quasi-Newton method. Then, we introduce the nonmonotone line search technique [14]; in such technique, some growth in the function value is permitted. The nonmonotone schemes can avoid being trapped in local minima, which improve the likelihood of finding a global optimum, and can also accelerate the convergence process. The third method presented in this section is the nonmonotone spectral gradient (NSG) method, which is a variant of "Global Barzilai–Borwein method" [23], such method combines the two previous methods (a nonmonotone line search strategy with the Barzilai and Borwein method). In Sect. 9.3 a review of the classical particle swarm optimization is provided. Detail description of the new hybrid method is presented in Sect. 9.4. In Sect. 9.5 numerical experiences are presented in the solution of some test problems. The chapter is concluded in Sect. 9.6.

9.2 Nonmonotone Spectral Gradient Method

9.2.1 *Iterative Search Method and Step Size*

It is well known that any solution of the unconstrained minimization problem

$$\min_{x \in \mathbb{R}^n} f(x) \tag{9.1}$$

where $f : \mathbb{R}^n \longrightarrow \mathbb{R}$ is a continuously differentiable function solves the nonlinear equations problem:

$$\text{find } x^\star \in \mathbb{R}^n \text{ such that } \nabla f(x^\star) = 0 \tag{9.2}$$

The methods proposed to solve it are usually iterative procedures: if x_k denotes the current iterate, and if it is not a good estimator of $x^\star$, a better one, $x_{k+1} = x_k + \alpha_k d_k$ is required. Here d_k is a search direction and $\alpha_k > 0$ is a steplength.

The search direction d_k is usually required to satisfy the descent condition

$$g_k^T d_k < 0 \tag{9.3}$$

The descent direction can be obtained by different methods, such as steepest descent method, Newton method, and quasi-Newton method. After the descent direction d_k is fixed, we need to choose a steplength α_k which will ensure a sufficient decrease of f.

In the line search methods, we compute a steplength that decides how far x_k should move along the direction d_k. The steplength can be determined either exactly or inexactly. In the exact line search method, we find α_k such that the objective function f in the direction d_k is minimized, i.e.,

$$f(x_k + \alpha_k d_k) = \min_{\alpha > 0} f(x_k + \alpha d_k) \tag{9.4}$$

Generally, this method is too expensive and ineffective; hence, it has been practically abandoned. Therefore the inexact line search methods are preferable. This class of methods can be divided into two classes:

- **The monotone line search method**: In this method (Armijo, Goldstein, Wolfe, and others) [2, 13, 26], one generates a limited number of trial steplength until it finds one that provides a sufficient decrease of the objective function. The Armijo rule, known also by the backtracking line search, is a good example of such line searches [2].
- **The nonmonotone line search methods**: In such technique, some growth in the function value is permitted, which gives rise to the term **nonmonotone**. Nonmonotone scheme can improve the likelihood of finding a global optimum, and can also accelerate the convergence process in cases where a monotone

scheme is forced to creep along the bottom of a narrow curved valley. The original nonmonotone line search strategy is proposed in [14].

9.2.2 Nonmonotone Spectral Gradient Methods

In 1988, Barzilai and Borwein [3] presented a new choice of step size for the gradient method for solving unconstrained minimization problems. Their method aimed to accelerate the convergence of the steepest descent method. The Barzilai–Borwein method, referred to (BB) algorithm, requires few storage locations and inexpensive computations. Therefore, several researchers have paid attention to it and have proposed some variants to solve large-scale unconstrained minimization problems. The convergence for quadratics was established by Raydan [22], and a global scheme was discussed more recently for nonquadratic functions [23] that uses a variant of the nonmonotone line search of Grippo, Lampariello, and Lucidi [14]. Other developments in BB algorithm can be found in [5, 10, 29].

9.2.2.1 The Barzilai–Borwein Method

The gradient iteration form

$$x_{k+1} = x_k - \alpha_k g_k \tag{9.5}$$

can be written as

$$x_{k+1} = x_k - A_k g_k \tag{9.6}$$

Where $A_k = \alpha_k I$. In order to make the matrix A_k have quasi-Newton property, we compute α_k such that

$$\min \|s_{k-1} - A_k y_{k-1}\| \tag{9.7}$$

This yields that

$$\alpha_k = \frac{s_{k-1}^T y_{k-1}}{y_{k-1}^T y_{k-1}} \tag{9.8}$$

By symmetry, we may minimize $\|A_k^{-1} s_{k-1} - y_{k-1}\|$ with respect to α_k and get

$$\alpha_k = \frac{s_{k-1}^T s_{k-1}}{s_{k-1}^T y_{k-1}} \tag{9.9}$$

Taking these ideas into account, the Barzilai–Borwein gradient algorithm which is known also as the spectral gradient (SG) algorithm is established in the following algorithm:

Algorithm (The Barzilai–Borwein Gradient Method [3, 22])

Step 0. $x_0 \in \mathbb{R}^n$, $\alpha_0 \in \mathbb{R}$, let $k = 0$.
Step 1. If $\|g(x_k)\| = 0$, stop, declaring that x_k is stationary.
Step 2. Let $d_k = -g_k$
Step 3. Compute α_k by (9.8) or (9.9)
Step 4. Set $x_{k+1} = x_k + \alpha_k d_k$.
Step 5. Let $k = k + 1$, return to **Step 1.**

The most important features of the previous algorithm are [23]:

- In this method no matrix computations and no line searches are required.
- Every iteration requires two inner products, one scalar–vector multiplication, two vector additions, and only one gradient evaluation.
- It is a gradient method which uses information of the two previous iterates. It makes a difference with the steepest descent method, which uses only information of the previous iterate.
- It satisfies the weak secant equation: $s_k^T A_{k+1} s_k = s_k^T y_k$.
- The scalar α_{k+1} is a Rayleigh quotient of the matrix

$$\int_0^1 \nabla^2 f(x_k + t s_k) dt$$

Barzilai and Borwein [3] proved that the above algorithm is R-superlinearly convergent for the quadratic case. However, Fletcher [11] argued that, in general, only R-linear convergence should be expected. Later, Raydan [22] established global convergence for the strictly convex quadratic case.

In the general nonquadratic case, a globalization strategy based on nonmonotone line search is suitable to Barzilai–Borwein gradient method [23].

9.2.2.2 The Nonmonotone Line Search Method

One possibility among several to ensure sufficient decrease of the objective function is the Armijo condition [2]:

$$f(x_k + \alpha_k d_k) \leq f(x_k) + \alpha_k \gamma d_k^T g_k \tag{9.10}$$

for some $\gamma \in (0, 1)$. The Armijo backtracking rule chooses the integer h_k minimal, such that (9.10) is satisfied for $\alpha_k = a\beta^{h_k}$, where $a > 0$, $\beta \in (0, 1)$.

The Armijo line search requires the function value to decrease monotonically at each iteration. As a result, it may cause the sequence of iterations following the bottom of a curved narrow valley, which commonly occurs in difficult nonlinear problems. To overcome this difficulty, a credible alternative is to allow an occasional increase in the objective function at each iteration. In 1986, Grippo, Lampariello, and Lucidi [14] proposed a nonmonotone generalization of the Armijo condition (9.10) for Newton's method, which was used in several subsequent papers and led to new nonmonotone algorithms.

In the nonmonotone line search, we will enforce a much weaker condition of the form

$$f(x_k + \alpha_k d_k) \leq \max_{0 \leq j \leq \min\{k,M\}} f(x_{k-j}) + \alpha_k \gamma d_k^T g_k \tag{9.11}$$

where M is a nonnegative integer and γ is a small positive number.

9.2.2.3 Nonmonotone Spectral Gradient (NSG) Algorithm

The condition (9.11) allows the objective function to increase at some iterations and still guarantees global convergence. This feature fits nicely with the nonmonotone behavior of the Barzilai and Borwein gradient method.

In [23], Raydan combined the Grippo–Lampariello–Lucidi nonmonotone line search [14] with the spectral gradient choice of steplength [3] to propose global Barzilai and Borwein (GBB) algorithm. This method requires only a few storage locations and inexpensive computations. Furthermore the global convergence method is guaranteed by using a nonmonotone line search strategy. The NSG algorithm [21], which is a variant of GBB algorithm [23], is presented in the following:

Algorithm (Nonmonotone Spectral Gradient)
Let $k = 0$.

Step 1. Detect whether the current point is stationary
If $\|g(x_k)\| = 0$, stop, declaring that x_k is stationary.
Step 2. Backtracking
Step 2.1 Compute $d_k = -\alpha_k g_k$. Set $\lambda = 1$.
Step 2.2 Set $\tilde{x} = x_k + \lambda d_k$.
Step 2.3 If

$$f(\tilde{x}) \leq \max_{0 \leq j \leq \min\{k,M\}} f(x_{k-j}) + \gamma \lambda \langle d_k, g_k \rangle \tag{9.12}$$

then set $\lambda_k = \lambda$, $\tilde{x}_k = \tilde{x}$ and go to **Step 3**,
else, define $\lambda_{new} \in [\sigma_1, \sigma_2 \lambda]$. Set $\lambda = \lambda_{new}$ and go to **Step 2.2**.
Step 3. Compute $b_k = \langle s_k, y_k \rangle$.
If $b_k \leq 0$, set $\alpha_{k+1} = \alpha_{\max}$, else, compute $a_k = \langle s_k, s_k \rangle$ and $\alpha_{k+1} = \min\{\alpha_{\max}, \max\{\alpha_{\min}, a_k/b_k\}\}$. **Step 4.** Let $k = k + 1$, return to **Step 1.**

Remark 9.1

1. The computation of λ_{new} uses one-dimensional quadratic interpolation [5, 6].
2. The (NSG) algorithm cannot cycle indefinitely between **Step 2.2** and **Step 2.3**, since $\lambda\langle d_k, g_k\rangle = -\lambda\alpha_k\|g_k\|^2 < 0$ and $\gamma < 1$, for sufficiently small value of λ the condition (9.12) is well defined.

9.3 Particle Swarm Optimization (PSO)

9.3.1 Metaheuristic Methods

In metaheuristic methods, one optimizes a problem by iteratively trying to improve a candidate solution with regard to a given measure of quality. Metaheuristics mainly invoke exploration and exploitation search procedures in order to diversify the search all over the search space and intensify the search in some promising areas. The advantage of metaheuristic methods is that they make few or no assumptions about the problem being optimized (no need for accurate guess values, derivatives, linearity, or a closed form) and can search very large spaces of candidate solutions. The disadvantage is that they give no exact solution (although by careful implementation, the results can be reproduced up to a given precision).

Some of the most popular metaheuristics are:

- The evolutionary algorithms, including: evolutionary strategies and genetic algorithms [15]
- simulated annealing [18],
- ant colonies algorithms [9],
- particle swarm optimization [16],
- tabu search [12].

The research is very active and it is impossible to produce an exhaustive list of different metaheuristic methods.

There are different ways to classify and describe metaheuristic algorithms. Depending on the characteristics selected to differentiate among them, several classifications are possible, each of them being the result of a specific viewpoint. A distinction can be made between single-individual methods and population-based methods. Algorithms working on single solutions are called trajectory methods and encompass local search-based metaheuristics, like tabu search and simulated annealing. They all share the property of describing a trajectory in the search space during the search process. Population-based metaheuristics, on the contrary, perform search processes which describe the evolution of a set of points in the search space. Evolutionary algorithms and particles swarm optimization are good examples of population-based algorithms.

9.3.2 Swarm Intelligence

Swarm intelligence (SI), which is an artificial intelligence (AI) discipline, is concerned with the design of intelligent multi-agent systems by taking inspiration from the collective behavior of social insects such as ants, termites, bees, and wasps, as well as from other animal societies such as flocks of birds or schools of fish.

Optimization techniques inspired by swarm intelligence have become increasingly popular during the last decade. The advantage of these approaches over traditional techniques is their robustness and flexibility. These properties make swarm intelligence a successful design paradigm for algorithms that deal with increasingly complex problems.

9.3.3 PSO as a Member of Swarm Intelligence

Particle swarm optimization (PSO) was introduced in 1995 by James Kennedy (social psychologist) and Russell Eberhart (electrical engineer) to simulate the natural swarming behavior of birds as they search for food [16].

The philosophy of PSO is based on the evolutionary cultural model, which states that in social environments individuals have two learning sources: individual learning and cultural transmission. Individual learning is an important feature in static and homogeneous environments because one individual can learn many things about the environment from a single interaction with it. However, if the environment is dynamic or heterogeneous, then that individual needs many interactions with the environment before it gets to know it. Because a single individual might not get enough chances to interact with such environment, cultural transmission (meaning learning from the experiences of others) becomes a requisite, too. In fact, individuals that have more chances to succeed in achieving their goals are the ones that combine both learning sources, thus increasing their gain in knowledge.

9.3.4 PSO Basic Algorithm

Suppose the following scenario: a swarm of birds is randomly searching food in an area. There is only one piece of food in the area being searched. All the birds do not know where the food is. But they know how far the food is in each iteration. So what is the best strategy to find the food? The effective one is to follow the bird, which is nearest to the food. PSO learned from the scenario and used it to solve the optimization problems. In PSO, each single solution is a "bird" in the search space. We call it "particle." All of particles have fitness values, which are evaluated by the fitness function to be optimized, and have velocities, which direct the flying of the particles. The particles fly through the problem space by following the current optimum particles.

Let n be the dimension of the search space and Ps the number of particles in the swarm, then $x_i = (x_{i1}, x_{i2}, \ldots, x_{in})$ denotes the position of the particle $i \in (1, 2, \ldots, Ps)$ of the swarm, and $p_i = (p_{i1}, p_{i2}, \ldots, p_{in})$ denotes the best position it has ever visited. The index of the best particle in the population (the one which has visited the global best position) is represented by the symbol g. At each iteration k in the simulation, the velocity of the ith particle, represented as $v_i = (v_{i1}, v_{i2}, \ldots, v_{in})$, is updated using the following equation:

$$v_{ij}^{k+1} = v_{ij}^k + c_1 r_1 (p_{ij}^k - x_{ij}^k) + c_2 r_2 (p_{gj}^k - x_{ij}^k) \tag{9.13}$$

where c_1 and c_2 are the acceleration constants, r_1 and r_2 are random numbers uniformly distributed in the interval [0, 1].

Each of the three terms of the velocity update equation (9.13) have different roles in the PSO method.

The first term v_i^k is responsible for keeping the particle moving in the same direction.

The second term $c_1 r_1 (p_{ij}^k - x_{ij}^k)$, called the **cognitive component**, acts as the particle's memory, causing it to tend to return to the regions of the search space in which it has experienced high individual fitness. The cognitive coefficient c_1 affects the size of the step the particle takes toward its individual best candidate solution p_i^k.

The third term $c_2 r_2 (p_{gj}^k - x_{ij}^k)$, called the **social component**, causes the particle to move to the best region the swarm has found so far. The social coefficient c_2 represents the size of the step, the particle takes toward the global best candidate solution p_g^k, the swarm has found up until that point.

Once the velocity for each particle is calculated, each particle's position is updated by applying the new velocity to the particle's previous position

$$x_i^{k+1} = x_i^k + v_i^{k+1} \tag{9.14}$$

This process is repeated until some stopping condition is met. Some common stopping conditions include: a preset number of iterations of the PSO algorithm, a number of iterations since the last update of the global best candidate solution, or a predefined target fitness value. Pseudocode for a PSO is shown below:

PSO Algorithm

1. Create the initial swarm: Random positions and velocities;
2. Evaluate the fitness of each particle;
3. **while** $k \le Kmax$
4.   **for** each particle $i = 1, \ldots, Ps$
5.   Update particle i according to equations (9.13) and (9.14);
6.   **if** $f(x_i) < f(p_i)$ **then**
7. $p_i = x_i$;

8. **if** $f(x_i) < f(p_g)$ **then**
9. $p_g = x_i$
10. **end if**
11. **end if**
12. **end for**
13. **end while**

9.3.5 *Modification to the Basic Algorithm*

Like many other metaheuristics, the PSO algorithm frequently faces the problem of being trapped in local optima. Balancing the global exploration (diversification) and local exploitation (intensification) abilities of PSO is therefore very important.

9.3.5.1 Velocity Bounds

Large velocities can cause particles to leave the defined boundary constraints of the problem, which not only results in wasted effort due to discarded solution, but may also cause the swarm to diverge rather than converge. To counter this problem the particle's velocity can be constricted to stay in a fixed range, by defining a maximum velocity value $V_{\max}$ and applying the following rule after every velocity updating:

$$v_{ij} \in [-V_{\max}, V_{\max}] \tag{9.15}$$

9.3.5.2 Inertia Weight

Motivated by the desire to provide balance between exploration and exploitation process and reduce the importance of $V_{\max}$, Shi and Eberhart came up with what they called PSO with inertia [24]. The inertia weight is multiplied by the previous velocity in the standard velocity equation. The following modification of the velocity update was proposed:

$$v_{ij}^{k+1} = \omega v_{ij}^{k} + c_1 r_1 (p_{ij}^{k} - x_{ij}^{k}) + c_2 r_2 (p_{gj}^{k} - x_{ij}^{k}) \tag{9.16}$$

The inertia weight determines the contribution rate of a particle's previous velocity to its velocity at the current time step. A large inertia weight facilitates a global search while a small inertia weight facilitates a local search. Therefore, the inertia weight must be adjusted for a better exploration–exploitation trade-off.

The performance of PSO has been greatly improved through experimental study, by introducing a linearly decreasing inertia weight into the original version of PSO [25, 27]. The linearly decreasing inertia weight ω decreases from ($\omega_{\max}$) to ($\omega_{\min}$) according to the following equation:

$$\omega_k = \omega_{\max} - \frac{(\omega_{\max} - \omega_{\min}) \times k}{K_{max}} \tag{9.17}$$

where K_{max} is the maximum number of iterations, and k is the current iteration.

Many other variations of inertia weight strategy have been proposed in the literature, a review of such variations is given chronologically in [4].

9.3.5.3 Constriction Factor

Another method for controlling the behavior of the particle swarm is the introduction of a constriction factor. Such a method was first proposed by Clerc and Kennedy in [8]. Velocities are constricted, with the following change in the velocity update:

$$v_{ij}^{k+1} = \chi(v_{ij}^k + \phi_1 r_1 (p_{ij}^k - x_{ij}^k) + \phi_2 r_2 (p_{gj}^k - x_{ij}^k)) \tag{9.18}$$

where

$$\chi = \frac{2}{|\phi - 2 + \sqrt{\phi^2 - 4\phi}|}, \phi = \phi_1 + \phi_2 > 4 \tag{9.19}$$

The value of ϕ is commonly set to 4.1 and the constant multiplier χ is approximately 0.7298.

By using the constriction coefficient, the amplitude of the particle's oscillation decreases, resulting in its convergence over time.

Note that a PSO with constriction is algebraically equivalent to a PSO with inertia. Indeed, (9.16) and (9.18) can be transformed into one another via the mapping $\chi \leftrightarrow \omega$, $c_1 \leftrightarrow \chi\phi_1$, and $c_2 \leftrightarrow \chi\phi_2$.

9.3.6 Neighborhood Topologies

In the original PSO, two different kinds of neighborhoods were defined for PSO:

- In the gbest swarm, all the particles are neighbors of each other; thus, the position of the best overall particle in the swarm is used in the social term of the velocity update equation. It is assumed that gbest swarms converge fast, as all the particles are attracted simultaneously to the best part of the search space. However, if the global optimum is not close to the best particle, it may be impossible to the swarm to explore other areas; this means that the swarm can be trapped in local optima.
- In the lbest swarm, only a specific number of particles (neighbor count) can affect the velocity of a given particle. The swarm will converge slower but can locate the global optimum with a greater chance.

In [17] a systematic review of alternative "social" neighborhood topologies was investigated.

9.4 The Proposed Hybrid Approach

This section describes the new method for the unconstrained minimization problem.

The best characteristics of PSO (strong global search ability), are combined with the good local search characteristics of the NSG, to develop a novel hybrid algorithm; the proposed algorithm is called NPSOG. The NPSOG method is used to find the global optimum.

Traditionally a hybrid PSO carries out first a certain number of iteration, and then an iterative method (gradient type methods or another metaheuristic) is applied to refine the approximations. On the contrary, in our new hybrid approach,[1] in every iteration of PSO, we perform a local search, by NSG method, around p_g, if two conditions are satisfied. First, the new solution x_i^k has to produce an objective value lower than the old one (for the minimization problems). Second, a randomly generated number has to be lower than the current corresponding loudness.

$$rand < A_i \qquad \& \qquad f(x_i^k) < f(x_i^{k-1}) \tag{9.20}$$

These conditions were also suggested in [7].

Note that the loudness parameter[2] A_i decreases as the particle gets closer to optimum.

The following algorithm is a formal description of the NPSOG method

NPSOG Algorithm

1. Create the initial swarm: Random positions and velocities;
2. Evaluate the fitness of each particle;
3. **while** $k \leq Kmax$
4.   **for** each particle $i = 1, \ldots, Ps$
5.   Update particle i according to equations (9.13) and (9.14);
6.   **if** $f(x_i) < f(p_i)$ **then**
7. $p_i = x_i$;
8.   **if** $f(x_i) < f(p_g)$ **then**
9. $p_g = x_i$
10.   **end if**
11.   **if** $(rand < A_i \ \& \ f(x_i^k) < f(x_i^{k-1}))$ **then**
12.   Local search, by NSG method, around p_g;
13.   Update p_g and reduce A_i
14.   **end if**
15.   **end if**
16.   **end for**
17. **end while**

---

[1]The idea of this work was inspired by [19, 20].

[2]The loudness parameter was introduced for the first time by Yang in his famous article about bat algorithm [28].

9.5 Simulation Results

In this section, we study the numerical behavior of the implementations in Scilab 5.5.2, obtained for a set of standard test problems, by means of the algorithms NPSOG, PSO, and CPSOG. Here CPSOG algorithm is a classical hybridization between PSO and NSG in which the PSO carries out first a certain number of iterations, and then the NSG method is applied to refine the approximations.

NSG Parameters
$\gamma = 10^{-4}$, $\alpha_{\min} = 10^{-30}$, $\alpha_{\max} = 10^{30}$, $\sigma_1 = 0.1$, $\sigma_2 = 0.9$, $M = 5$, the maximum number of iteration is $Nmax = 100$, and $\alpha_0 = 1/\|\nabla f(x_0)\|_\infty$, where x_0 is the initial guess.

PSO Parameters

1. Population size = 100.
2. Maximum number of iterations $Kmax = 100$.
3. Acceleration constants $c_1 = c_2 = 1.7$.
4. The linearly decreasing method is adopted for the inertia weight, with ($\omega_{\max} = 1.4$ $\omega_{\min} = 0.4$)
5. fitness function = the objective function.

The Loudness Parameter
The loudness parameter was updated as follows [7]:

$$A_i = \frac{A_0 - A_\infty}{1 - Kmax}(k - Kmax) + A_\infty \tag{9.21}$$

where the index 0 and ∞ stand for the initial and final values, respectively ($A_0 = 0.9$, $A_\infty = 0.6$).

Below, we list the test functions solved by the algorithms. These test functions have different features, with known global optima [1].

1. **Extended Powell singular quartic function**
   $f_1 = \sum_{i=1}^{n/4} [(x_{4i-1} + 10x_{4i-2})^2 + 5(x_{4i-1} - x_{4i})^2 + (x_{4i-2} + 2x_{4i-1})^4 + (x_{4i-3} + 10x_{4i})^4]$,
   $x_i \in [-100, 100]$, $f_1(x^\star) = 0$.
2. **Goldstein-Price's function**
   $f_2 = [1 + (x_1 + x_2 + 1)^2(19 - 14x_1 + 3x_1^2 - 14x_2 + 6x_1x_2 + 3x_2^2)][30 + (2x_1 - 3x_2)^2(18 - 32x_1 + 12x_1^2 + 48x_2 - 36x_1x_2 + 27x_2^2)]$,
   $x_i \in [-2, 2]$, $f_2(x^\star) = 3$.
3. **Shubert function**
   $f_3 = \sum_{j=1}^{5} j\cos((j+1)x_1 + j) \sum_{j=1}^{5} j\cos((j+1)x_2 + j)$,
   $x_i \in [-50, 50]$, $f_3(x^\star) = -186.7309$

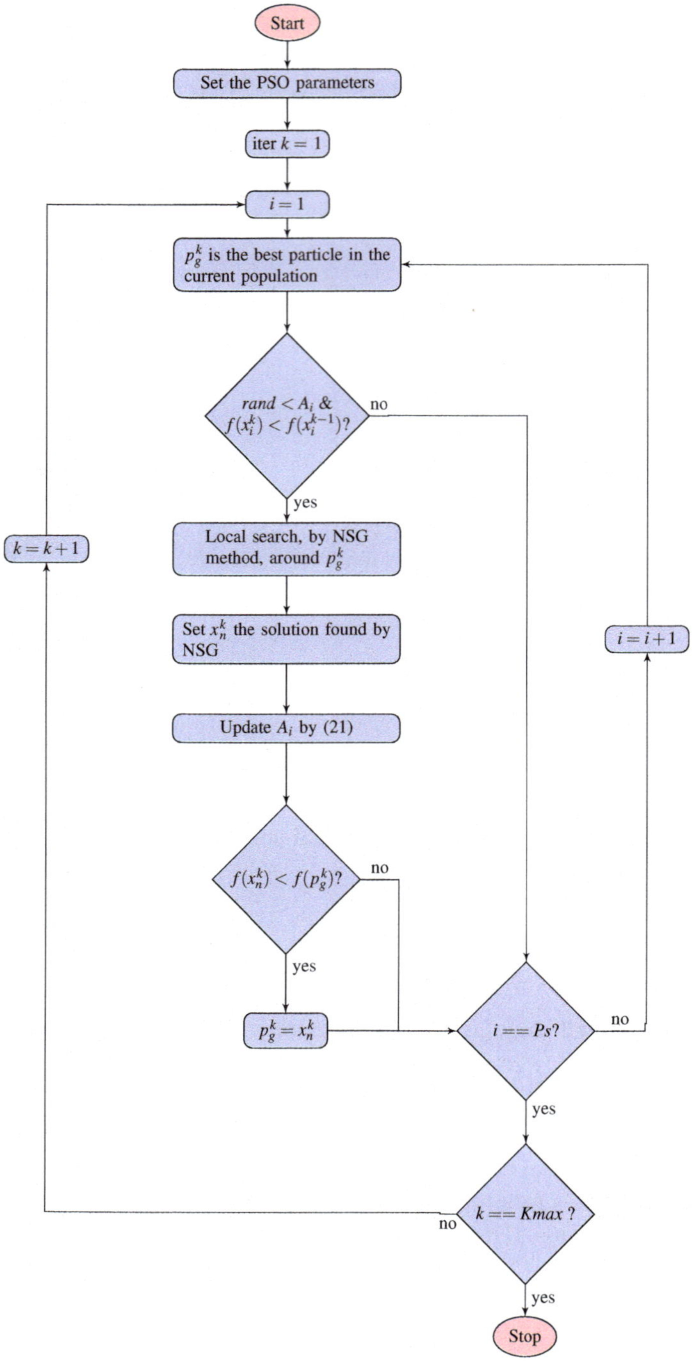

Fig. 9.1 Flow chart of the NPSOG algorithm

4. **Rastrigin function**
   $f_4 = x_1^2 + x_2^2 - 10\cos(2\pi x_1) - 10\cos(2\pi x_2) + 20$
   $x_i \in [-2, 2],\ f_4(x^\star) = 0.$
5. **Function f_5**
   $f_5 = \sin^2(x_1^2 + x_2^2) + x_1^2 + x_2^2,$
   $x_i \in [-50, 50],\ f_5(x^\star) = 0.$

In the following tables we report the smallest value in the objective function reached by the algorithms for 10 runs, the average of minimums and the standard deviation[3] (SD).

Note that the Standard Deviation is defined by

$$SD = \sqrt{\frac{1}{N}\sum_{i=1}^{N}(e_i - e^\star)^2}$$

where $e^\star$ is the optimal solution, e_i is the solution from the ith run, and N is the number of runs.

9.6 Conclusion

The numerical results of Tables 9.1, 9.2, 9.3 show that, in general, the smallest value in the objective function, the average of minimums and the standard deviation given by NPSOG are significantly smaller than those given by CPSOG.

Note that the results given by the two hybrid approach and the results NPSOG and CPSOG are more better than those given by PSO algorithm because of the use of the local search by NSG method.

Table 9.1 Comparison between the smallest value in the objective functions

$F(n)$[a]	PSO	CPSOG	NPSOG
$f_1(4)$	60.31219	0.0000227	$6.236e-10$
$f_1(40)$	$6.152e+08$	25.423383	$3.219e-12$
$f_1(100)$	$1.122e+10$	403.40584	$7.712e-11$
$f_2(2)$	6.0209915	3	3
$f_3(2)$	-182.3714	-182.3714	-182.3714
$f_4(2)$	1.0540639	0	0
$f_5(2)$	0.9867069	$4.413e-20$	$4.413e-20$

[a] n is the problem dimension

[3]The standard deviation is used to indicate the stability of an algorithm, a more stable algorithm should produce a smaller value of this measurement.

Table 9.2 Comparison between the average of minimums

$F(n)$	PSO	CPSOG	NPSOG
$f_1(4)$	487146.39	7.6065385	$1.423e-08$
$f_1(40)$	$3.179e+09$	2297.8701	0.0000002
$f_1(100)$	$1.605e+10$	11346.13	0.0000001
$f_2(2)$	28.689424	11.1	3
$f_3(2)$	-113.02962	-154.36692	-186.73091
$f_4(2)$	4.5429155	1.4924381	$2.487e-15$
$f_5(2)$	33.649999	$3.727e-12$	$1.829e-12$

Table 9.3 Comparison between the standard deviations

$F(n)$	PSO	CPSOG	NPSOG
$f_1(4)$	666877.15	14.679889	$1.956e-08$
$f_1(40)$	$3.476e+09$	3924.3487	0.0000002
$f_1(100)$	$1.622e+10$	13593.779	0.0000002
$f_2(2)$	33.274943	14.788509	$5.217e-14$
$f_3(2)$	88.243807	54.743889	0.0000088
$f_4(2)$	4.9680153	1.8074318	$5.838e-15$
$f_5(2)$	39.456341	$8.285e-12$	$5.755e-12$

We conclude that our proposed method seems to be an interesting candidate for solving unconstrained differentiable optimization.

References

1. Andrei, N.: An unconstrained optimization test functions collection. Adv. Model. Optim. **10**(1), 147–161 (2008)
2. Armijo, L.: Minimization of functions having Lipschitz first partial derivatives. Pac. J. Math. **16**(1), 1–3 (1966)
3. Barzilai, J., Borwein, J.M.: Two point step size gradient methods. IMA J. Numer. Anal. **8**, 141–148 (1988)
4. Bansal, J.C., Singh, P.K., Saraswat, M., Verma, A., Jadon, S.S., Abraham, A.: Inertia Weight Strategies in Particle Swarm Optimization, pp. 633–640. Third World Congress on Nature and Biologically Inspired Computing, Salamanca (2011)
5. Birgin, E.G., Martínez, J.M., Raydan, M.: Nonmonotone Spectral Projected Gradient Methods on Convex Sets. SIAM J Optim. **10**(4), 1196–1211 (2000)
6. Birgin, E.G., Raydan, M.: SPG software for convex-constrained optimization. ACM Trans. Math. Soft. **27**(3), 340–349 (2001)
7. Chakri, A., Khelif, R., Benouaret, M., Yang, X.Y.: New directional bat algorithm for continuous optimization problems. Expert Syst. Appl. **69**, 159–175 (2017)
8. Clerc, M., Kennedy, J.: The particle swarm - explosion, stability, and convergence in a multidimensional complex space. IEEE T Evol. Comput. **6**(1), 58–73 (2002)
9. Colorni, A., Dorigo, M., Maniezzo, V.: An investigation of some proprieties of an "ant algorithm". In: Proc. Parallel Problem Solving from Nature Conference, pp. 509–520. Elsevier Publishing (1992)

10. Dai, Y.H., Hager, W.W., Schitkowski, K., Zhang.: The cyclic Barzilai-Borwein method for unconstrained optimization. IMA J. Numer. Anal. **26**(3), 1–24 (2006)
11. Fletcher, R.: Low storage methods for unconstrained optimization. In: Lect. Appl. Math., vol. 26, pp. 165–179. American Mathematical Society, Providence, RI (1990)
12. Glover, F.: Future paths for integer programming and links to artificial intelligence. Comput. Oper. Res. **13**(5), 533–549 (1986)
13. Goldstein, A.A.: On steepest descent. SIAM J. Control. **3**, 147–151 (1965)
14. Grippo, L., Lampariello, F., Lucidi, S.: A nonmonotone line search technique for Newton's method. SIAM J. Numer. Anal. **23**, 707–716 (1986)
15. Holland, J.H.: Adaptation in Natural and Artificial Systems: An Introductory Analysis with Applications to Biology, Control, and Artificial Intelligence. University Michigan Press, Oxford, England (1975)
16. Kennedy, J., Eberhart, R.: Particle Swarm Optimization. In: IEEE Int. Conf. Neural Networks, Perth, WA, Australia, vol. 4, pp. 1942–1948 (1995)
17. Kennedy, J., Mendes, R.: Population structure and particle swarm performance. In: Proc. Congress on Evolutionary Computation (CEC'02), vol. 2, pp. 1671–1676 (2002)
18. Kirkpatrick, S., Gelatt, C.D., Vecchi, M.P.: Optimization by simulated annealing. Science, New Series **220**(4598), 671–680 (1983)
19. La Cruz, W., Noguera, N.: Hybrid spectral gradient method for the unconstrained minimization problem. J. Global Optim. **44**, 193–212 (2009)
20. Lakhbab, H., El Bernoussi, S.: Hybrid nonmonotone spectral gradient method for the unconstrained minimization problem. Comput. Appl. Math. **36**(3), 1421–1430 (2017)
21. Lakhbab, H., El Bernoussi, S.: A hybrid method based on particle swarm optimization and nonmonotone spectral gradient method for unconstrained optimization problem. Int. J. Math. Anal. **6**(60), 2963–2976 (2012)
22. Raydan, M.: On the Barzilai and Borwein choice of steplength for the gradient method. IMA J. Numer. Anal. **13**, 321–326 (1993)
23. Raydan, M.: The Barzilai and Borwein gradient method for the large scale unconstrained minimization problem. SIAM J. Optim. **7**(1), 26–33 (1997)
24. Shi, Y., Eberhart, R.C.: A modified particle swarm optimizer. In: Proc. IEEE Inter. Conf. on Evolutionary Computation, Anchorage, AK, USA, pp. 69–73 (1998)
25. Shi, Y., Eberhart, R.C.: Experimental study of particle swarm optimization. In: Proc. 1999 Congress on Evol. Comput-CEC99, Washington, DC, USA, vol. 3, pp. 1945–1950 (1999)
26. Wolfe, P.: Convergence conditions for ascent methods. SIAM Rev. **11**(2), 226–235 (1969)
27. Xin, J., Chen, G., Hai, Y.: A particle swarm optimizer with multistage linearly-decreasing inertia weight. IEEE Inter. Conf. Comput. Sci. Opt. CSO 2009. **1**, 505–508 (2009)
28. Yang, X.S.: A new metaheuristic bat-inspired algorithm. In: González, J.R., Pelta, D.A., Cruz, C., Terrazas, G., Krasnogor, N. (eds) Nature Inspired Cooperative Strategies for Optimization. Studies in Computational Intelligence, vol. 284. Springer, Berlin, Heidelberg (2010)
29. Zhang, Y., Sun, W., Qi, L.: A nonmonotone filter Barzilai-Borwein method for optimization. Asia Pac. J. Oper. Res. **27**(1), 55–69 (2010)

Chapter 10
Detection of HIV-1 Protease Cleavage Sites via Hidden Markov Model and Physicochemical Properties of Amino Acids

Elif Doğan Dar, Vilda Purutçuoğlu, and Eda Purutçuoğlu

10.1 Introduction

AIDS(acquired immunodeficiency syndrome) is a disease that weakens the immune system by reducing the T-cells in the body which fight off infections [28]. In 2017, around 37 million people globally were living with HIV making 77 million since the start of the epidemic. Furthermore, approximately 1 million people died from AIDS-related illnesses in 2017 making a total of 35 million people since the start of the epidemic [35]. HIV-1 (human immunodeficiency virus-1) is the virus which causes AIDS gradually [7]. However, HIV-1 protease enzyme is needed for HIV-1 to be active. It cleaves newly synthesized polyproteins of the host cell to create the mature protein components that a HIV virion requires [15]. Usually, this enzyme extends to 8 amino acids long octamer sites on the polyprotein to cleave between the 4th and 5th amino acids [18]. Occasionally, these cleavage sites can also be heptamers or nonamers. HIV-1 protease inhibitors are good candidates for the AIDS treatment. Therefore, learning the key and the lock relationship between the enzyme and cleavage sites are crucial to finding the proper inhibitor key which locks the enzyme and prohibits it to create new active proteins for the virion. On the other hand, it is impossible to experimentally test all possible cleavage sites. Because there are 20 amino acids possible at each position, resulting in $20^8 = 2.56 \times 10^{10}$ cleavage sides. Hence, several methods have been used in the literature to predict the cleavage sites given earlier experimentally checked data.

E. Doğan Dar · V. Purutçuoğlu (✉)
METU, Middle East Technical University, Department of Statistics, Ankara, Turkey
e-mail: elif.dar@metu.edu.tr; vpurutcu@metu.edu.tr

E. Purutçuoğlu
University of Health Sciences, Department of Social Work, Ankara, Turkey

© Springer Nature Switzerland AG 2020

J. A. T. Machado et al. (eds.), *Numerical Solutions of Realistic Nonlinear Phenomena*, Nonlinear Systems and Complexity 31, https://doi.org/10.1007/978-3-030-37141-8_10

In the literature, some well-known approaches are used to predict HIV-1 cleavage sites including support vector machines(SVMs) [3], artificial neural networks (ANNs) [2, 33], and different encoding techniques [34], such as orthonormal encoding (OE) [2, 23]. In addition, the physicochemical properties of amino acids are being used in many papers. For instance, Jaeger et al. [10] use 4 biophysical properties, namely, hydropathy index, molecular mass, polarity, and occurrence percentage and Kim et al. [14] suggest a feature subset selection method using multi-layered perceptron (MLP) learning. Also, some researchers perform a subset of physicochemical properties from the AAIndex database [24, 34]. For a comprehensive review on the HIV-1 cleavage site detection, Rögnvaldsson [27] can be also seen.

Hereby, the present study proposes a hidden Markov model to capture the sequential nature of the problem. Hidden Markov model (HMM) is a model which utilizes the sequential relationship of the data unlike many other methods above. HMM is successfully applied to speech [13], handwriting [12], and gesture recognition [30] as well as biological applications such as the sequence alignment [5, 25], gene prediction [20], and the protein modeling [31, 36]. HMM is also implemented to the HIV-1 cleavage site detection problem [11]. However, in this model, random starting parameters are given to HMM instead of a guided choice. In the present study, guided starting parameters using physicochemical properties of the amino acids from the AAIndex database [22] are proposed inspired by the work of Zhang et al. [37]. Indeed, the physicochemical properties have been used earlier in the HIV-1 protease research [24, 34] but not in conjunction with HMM.

Hence, there are 544 features for each amino acid in this database. Since many machine learning algorithms suffer from the high dimensionality of the feature set, we implement a clustering based feature selection approach in our analyses [4, 19]. Here, we apply different strategies. Initially, we examine the k-means clustering and the hierarchical clustering methods based on the correlation measures and then, we randomly choose a representative from each cluster. Also, we perform the k-medoids clustering and select medoids as the cluster representatives. Later, we create hidden states including amino acids using chosen features. On the other side, since an amino acid can be grouped in multiple ways according to common features, it can share with groups. Due to this fact, the nature of the problem can be considered under a fuzzy clustering [1, 8, 17] and in this study, we use these methods too. Finally, by using the proposed approaches, we show that the states which we create give better results than earlier states suggested by Zhang et al. [37]. In that paper, they built a graph using the number of common features that amino acids share and declare the cliques of this graph as states.

Thus, in the organization of the study, we present HMM and its inference via a toy dataset in Sect. 10.2. In Sect. 10.3, we represent the application in a real dataset and discuss the outputs. Lastly, in Sect. 10.4, we conclude the outputs and present our future work by discussing both its mathematical and social aspects.

10.2 Hidden Markov Model

The hidden Markov model(HMM) is a special case of the probabilistic graphical models where entities are represented by nodes and dependencies by edges between them. In general, the probabilistic graphical models are intractable. However, in HMMs, most of the dependencies are replaced with independence relations. These assumptions help to make the problem tractable while keeping the necessary spatial dependencies. HMM has a sequence of observations and a sequence of states which produces them. We denote the observation sequence as $O = (O_1, O_2, \ldots, O_T)$ where each observation is an object from the set $o = \{o_1, o_2, \ldots, o_M\}$. Here, T represents the length of the observation and state sequences, and M denotes the number of possible observations. The hidden states which produce these observations are shown by $S = (S_1, S_2, \ldots, S_T)$ where each state is an object from a set of states $s = \{s_1, s_2, \ldots, s_N\}$. Here, N represents the number of possible states. We also define the following conditional independence assumptions.

- $P(O_k|S_1, \ldots, S_T, O_1, \ldots, O_T) = P(O_k|S_k)$ for any $1 \leq k \leq T$.
- $P(O_i, O_j|S_i, S_j) = P(O_i|S_i, S_j)P(O_j|S_i, S_j) = P(O_i|S_i)P(O_j|S_j)$ for $1 \leq i, j \leq T$.
- $P(S_k|S_1, \ldots, S_{k-1}) = P(S_k|S_{k-1})$ for any $2 \leq k \leq T$, i.e., states form a Markov chain.

Because of these assumptions, the joint probability of the system can be written as

$$\begin{aligned} P(O_1, \ldots, O_T, S_1, \ldots, S_T) &= P(O_1, \ldots, O_T|S_1, \ldots, S_T)P(S_1, \ldots, S_T) \\ &= P(O_1|S_1)P(O_2|S_2)\ldots P(O_T|S_T)P(S_1)P(S_2|S_1) \\ &\quad P(S_3|S_2)\ldots P(S_T|S_{T-1}) \\ &= \left(\prod_{i=1}^{T} P(O_i|S_i)\right) P(S_1) \left(\prod_{i=2}^{T} P(S_i|S_{i-1})\right). \end{aligned} \tag{10.1}$$

Therefore, to define an HMM we only need the probabilities below.

- Transition probabilities: $a_{ij} = P(S_t = j|S_{t-1} = i)$ for $1 \leq i, j \leq N$.
- Emission probabilities: $b_{ij} = P(O_t = j|S_t = i)$ for $1 \leq i \leq N$ and $1 \leq j \leq M$.
- Initial probabilities: $\pi_i = P(S_1 = s_i)$ for $1 \leq i \leq N$.

Furthermore, we can write transition and emission probabilities in a matrix form, say A and B and initial probabilities as a vector Π. Hereby, we denote parameters for HMM as $\lambda = (A, B, \Pi)$. In modeling via HMM, we are interested in basically solving three problems:

- Finding likelihood of an observation sequence given a HMM with parameters λ,
- Finding the most probable state sequence given the model parameters and the observation sequence,
- Estimating the model parameters given sequences of states and observations.

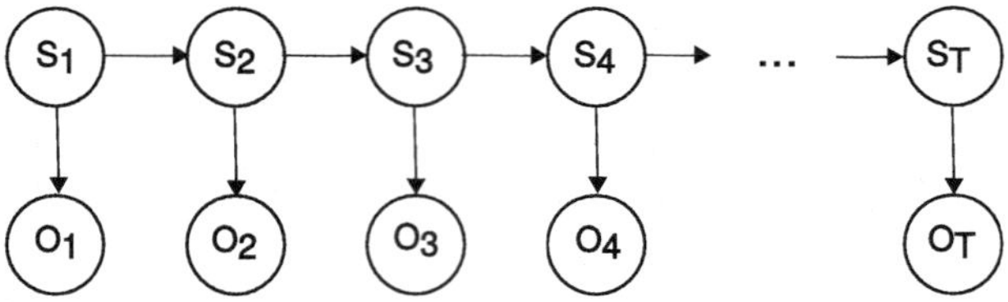

Fig. 10.1 Hidden Markov model; nodes represent hidden states and observations while edges indicate the dependencies among them

In the following part, we represent each step in detail by using a toy example whose description is also presented (Fig. 10.1).

10.2.1 Toy Example

For illustrative purposes, we use the following example which is modified from the Eisner's paper [6]. Let us say the number of ice creams that a person eats every day depends on the weather, which will be taken as either cold or hot. Also, let the number of ice creams she/he eats be from the set $\{1, 2, 3\}$. Here, the weather is the hidden variable where the number of ice creams is observed. Therefore, we have $S = \{H, C\}$ and $O = \{1, 2, 3\}$. Accordingly, the parameters of the model are defined as below:

$$A = \begin{bmatrix} P(S_t = H|S_{t-1} = H) & P(S_t = C|S_{t-1} = H) \\ P(S_t = H|S_{t-1} = C) & P(S_t = C|S_{t-1} = C) \end{bmatrix} = \begin{bmatrix} 0.6 & 0.3 \\ 0.4 & 0.5 \end{bmatrix},$$

$$B = \begin{bmatrix} P(O_t = 1|S_t = H) & P(O_t = 2|S_t = H) & P(O_t = 3|S_t = H) \\ P(O_t = 1|S_t = C) & P(O_t = 2|S_t = C) & P(O_t = 3|S_t = C) \end{bmatrix} = \begin{bmatrix} 0.2 & 0.4 & 0.4 \\ 0.5 & 0.4 & 0.1 \end{bmatrix},$$

$$\Pi = \begin{bmatrix} P(S_1 = H) \\ P(S_1 = C) \end{bmatrix} = \begin{bmatrix} 0.8 \\ 0.2 \end{bmatrix}.$$

We use this example to elaborate the calculations of the HMM steps.

10.2.2 Calculation of Likelihood

In some problems, we might be interested in finding the likelihood of an observation sequence given the model parameters while the state sequence is hidden. There are three major approaches for this calculation. We firstly describe the most natural way to solve this problem, which is the naive approach, and continue with faster counterparts: forward and backward algorithms.

10.2.2.1 Naive Approach

In this computation, we initially find the likelihood given a specific state as shown earlier and then, we sum over all possible states as below.

$$P(O|\lambda) = \sum_{S} P(O, S|\lambda), \tag{10.2}$$

where λ is the model parameter. In Eq. (10.2) there are N^T possible states. Therefore, when N and T are large, this approach becomes computationally demanding in the order of $\mathcal{O}(N^T)$ to calculate the likelihood. If we apply this on our toy example, assuming the observations for 3 days to be $O = (2, 1, 3)$, the likelihood of this observation sequence is found by

$$P(O = (2, 1, 3)|\lambda) = \sum_{S} P(O = (2, 1, 3), S = (s_1, s_2, s_3)|\lambda). \tag{10.3}$$

To find the sum in Eq. (10.3), let us first calculate one specific element in the sum, such as $S = (H, H, C)$, by using Eq. (10.1). Here, we obtain

$$\begin{aligned} P(O = (2, 1, 3), S = (H, H, C)|\lambda) &= P(O_1|S_1)P(O_2|S_2)P(O_3|S_3)P(S_1)P(S_2|S_1)P(S_3|S_2) \\ &= P(2|H)P(1|H)P(3|C)P(H)P(H|H)P(C|H) \\ &= 0.4 \times 0.2 \times 0.1 \times 0.8 \times 0.6 \times 0.3. \end{aligned}$$

We have to repeat this calculation $2^3 = 8$ times for different state sequences. Then, we need to compute their sum in order to obtain the likelihood. But, in real life examples, the number of state sequences can be very high. Thereby, another method which is faster than this naive approach is necessary.

10.2.2.2 Forward Algorithm

The forward algorithm [16] is a dynamic programming example where we break the problem into subproblems and use the earlier results in a recursion. In this way, we can solve the inference problem faster than the naive approach. Hereby, the likelihood of the observation sequence and a specific state at the last position of the state sequence given model parameters summed over all possible states are presented as below.

$$P(O|\lambda) = P_\lambda(O) = \sum_{i=1}^{N} P(S_T = s_i, O|\lambda). \tag{10.4}$$

Thus, in order to find the term in the summation conditional on the model parameters λ, we define

$$\alpha_k(S_k) = P_\lambda(S_k = s_i, O_1, \ldots, O_k). \tag{10.5}$$

The value in the sum is simply equal to $\alpha_T(s_i)$. To be able to find this term in Eq. (10.5), we can write it recursively via

$$\begin{aligned}
\alpha_k(S_k) &= \sum_{S_{k-1}=s_1}^{s_N} P_\lambda(S_k, S_{k-1}, O_1, \ldots, O_k) \\
&= \sum_{S_{k-1}=s_1}^{s_N} P_\lambda(O_k|S_k, S_{k-1}, O_1, \ldots, O_{k-1}) P_\lambda(S_k|S_{k-1}, O_1, \ldots, O_{k-1}) P_\lambda(S_{k-1}, O_1, \ldots, O_{k-1}) \\
&= \sum_{S_{k-1}=s_1}^{s_N} P_\lambda(O_k|S_k) P_\lambda(S_k|S_{k-1}) P_\lambda(S_{k-1}, O_1, \ldots, O_{k-1}) \\
&= \sum_{S_{k-1}=s_1}^{s_N} b_{s_k,o_k} a_{s_{k-1},s_k} \alpha_{k-1}(S_{k-1})
\end{aligned} \tag{10.6}$$

for $2 \leq k \leq T$, and for $k = 1$ we have

$$\alpha_1(S_1) = P_\lambda(S_1, O_1) = P_\lambda(S_1) P_\lambda(O_1|S_1) = \Pi(S_1) b_{S_1,O_1}. \tag{10.7}$$

Here, we recursively find $\alpha_1(S_1)$,...,$\alpha_T(S_T)$ and sum $\alpha_T(S_T)$ over all possible values of S_T to get the likelihood of interest. We complete the forward algorithm with the complexity $\mathcal{O}(N^2T)$. For large values of N and T, this complexity is lower than the complexity of the naive approach. Accordingly, let us see how the algorithm works on our toy example for $O = (2, 1, 3)$:

$$\alpha_1(S_1) = P_\lambda(S_1) P_\lambda(O_1|S_1) = \begin{cases} P_\lambda(H) P_\lambda(2|H) & \text{for } s_1 = H \\ P_\lambda(C) P_\lambda(2|C) & \text{for } s_1 = C \end{cases},$$

$$\begin{aligned}
\alpha_2(S_2) &= \sum_{S_1=s_1}^{s_N} P_\lambda(O_2|S_2) P_\lambda(S_2|S_1) \alpha_1(S_1) \\
&= P_\lambda(1|S_2) P_\lambda(S_2|H) \alpha_1(H) + P_\lambda(1|S_2) P_\lambda(S_2|C) \alpha_1(C) \\
&= \begin{cases} P_\lambda(1|H) P_\lambda(H|H) \alpha_1(H) + P_\lambda(1|H) P_\lambda(H|C) \alpha_1(C) & \text{for } s_2 = H \\ P_\lambda(1|C) P_\lambda(C|H) \alpha_1(H) + P_\lambda(1|C) P_\lambda(C|C) \alpha_1(C) & \text{for } s_2 = C \end{cases},
\end{aligned}$$

$$
\begin{aligned}
\alpha_3(S_3) &= \sum_{S_2=s_1}^{s_N} P_\lambda(O_3|S_3)P_\lambda(S_3|S_2)\alpha_2(S_2) \\
&= P_\lambda(3|S_3)P_\lambda(S_3|H)\alpha_2(H) + P_\lambda(3|S_3)P_\lambda(S_3|C)\alpha_2(C) \\
&= \begin{cases} P_\lambda(3|H)P_\lambda(H|H)\alpha_2(H) + P_\lambda(3|H)P_\lambda(H|C)\alpha_2(C) & \text{for } s_3 = H \\ P_\lambda(3|C)P_\lambda(C|H)\alpha_2(H) + P_\lambda(3|C)P_\lambda(C|C)\alpha_2(C) & \text{for } s_3 = C \end{cases}.
\end{aligned}
$$

Thus, finally we can obtain

$$
P_\lambda(O = (2, 1, 3)) = \alpha_3(H) + \alpha_3(C). \tag{10.8}
$$

10.2.2.3 Backward Algorithm

The Backward algorithm [16] is similar to the forward algorithm, except the starting point of the calculation. Hereby, we find the likelihood by the following expression.

$$
\begin{aligned}
P_\lambda(O) &= \sum_{i=1}^{N} P_\lambda(S_1 = s_i, O) \\
&= \sum_{i=1}^{N} P_\lambda(S_1 = s_i)P_\lambda(O_1|O_2, \ldots, O_T, S_1 = s_i)P_\lambda(O_2, \ldots, O_T|S_1 = s_i) \\
&= \sum_{i=1}^{N} P_\lambda(S_1 = s_i)P_\lambda(O_1|S_1 = s_i)P_\lambda(O_2, \ldots, O_T|S_1 = s_i) \\
&= \sum_{i=1}^{N} \Pi(s_i)b_{s_i,O_1} P_\lambda(O_2, \ldots, O_T|S_1 = s_i).
\end{aligned} \tag{10.9}
$$

To obtain the solution, we need to obtain the last term in the sum and we compute it by recursively using the following definition.

$$
\begin{aligned}
\beta_k(S_k) &= P_\lambda(O_{k+1}, \ldots, O_N|S_k) \qquad (10.10) \\
&= \sum_{S_{k+1}=s_1}^{s_N} P_\lambda(O_{k+1}, \ldots, O_T, S_{k+1}|S_k) \\
&= \sum_{S_{k+1}=s_1}^{s_N} P_\lambda(O_{k+2}, \ldots, O_T|S_{k+1}, S_k, O_{K+1})P_\lambda(O_{k+1}|S_{k+1}, S_k)P_\lambda(S_{k+1}|S_k) \\
&= \sum_{S_{k+1}=s_1}^{s_N} P_\lambda(O_{k+2}, \ldots, O_T|S_{k+1})P_\lambda(O_{k+1}|S_{k+1})P_\lambda(S_{k+1}|S_k)
\end{aligned}
$$

$$= \sum_{S_{k+1}=s_1}^{s_N} \beta_{k+1}(S_{k+1}) b_{S_{k+1},O_{k+1}} a_{S_k,S_{k+1}} \tag{10.11}$$

for $1 \le k \le N-1$. For $\beta_T(S_T)$, we cannot use the above definition since it involves O_{N+1} which does not exist. So, if we use our recursion formula for $k = T-1$,

$$\begin{aligned}\beta_{T-1}(S_{T-1}) &= \sum_{S_T=s_1}^{s_N} P_\lambda(O_T, S_T|S_{T-1}) \\ &= \sum_{S_T=s_1}^{s_N} \beta_T(S_T) P_\lambda(O_T|S_T) P_\lambda(S_T|S_{T-1}).\end{aligned} \tag{10.12}$$

But $P_\lambda(O_T, S_T|S_{T-1})$ can be also written as,

$$\begin{aligned}P_\lambda(O_T, S_T|S_{T-1}) &= P_\lambda(O_T|S_T, S_{T-1}) P_\lambda(S_T|S_{T-1}) \\ &= P_\lambda(O_T|S_T) P_\lambda(S_T|S_{T-1}).\end{aligned} \tag{10.13}$$

Therefore, for Eq. (10.13) to hold, $\beta_T(S_T) = 1$. Now by using Eq. (10.1) and the definition of β, we can get

$$P_\lambda(O) = \sum_{i=1}^{N} \Pi(s_i) b_{s_i,O_1} \beta_1(S_1 = s_i). \tag{10.14}$$

10.2.3 Viterbi Algorithm: Inference of the Most Probable Path

The Viterbi algorithm [16] is a recursive algorithm that is used to find the most probable sequence, also called *path*, given the observation sequence and parameters. In the calculation, after initialization of the state, at each step, we use the earlier paths which we find. More formally, our aim is to find

$$S^* = \arg\max_S P(S|O). \tag{10.15}$$

Note that:
If $f(a) \ge 0$ for all a and $g(a, b) \ge 0$ for all a, b, we have

$$\max_{a,b} f(a)g(a, b) = \max_a \left\{ f(a) \max_b g(a, b) \right\}, \tag{10.16}$$

and we have

$$\arg\max_S P(S|O) = \arg\max_S P(S, O) \tag{10.17}$$

since $P(O)$ does not contain any element from hidden states. Now let us define the function μ and the recursion by using Eq. (10.16) as below.

$$\begin{aligned}
\mu_k(S_k) &= \max_{S_1,\ldots,S_{k-1}} P(S_1, \ldots, S_k, O_1, \ldots, O_k) \\
&= \max_{S_1,\ldots,S_{k-1}} P(O_k|S_k)P(S_k|S_{k-1})P(S_1, \ldots, S_{k-1}, O_1, \ldots, O_{k-1}) \\
&= \max_{S_{k-1}} P(O_k|S_k)P(S_k|S_{k-1}) \max_{S_1,\ldots,S_{k-2}} P(S_1, \ldots, S_{k-1}, O_1, \ldots, O_{k-1}) \\
&= \max_{S_{k-1}} P(O_k|S_k)P(S_k|S_{k-1})\mu_{k-1}(S_{k-1})
\end{aligned} \tag{10.18}$$

for $2 \le k \le T$, and by definition $\mu_1(S_1) = P(S_1, O_1) = P(S_1)P(O_1|S_1)$. So, we find the sequence of the state which leads to

$$\max_{S_T} \mu_T(S_T) = \max_{S_1,\ldots,S_T} P(S_1, \ldots, S_T, O_1, \ldots, O_T). \tag{10.19}$$

For this purpose, at each iteration we note the most probable state and the path which satisfy these conditions. We can explain the application of this searching process via our toy example. Let $O = (2, 1, 3)$, then,

$$\mu_1 S_1 = P(S_1)P(O_1|S_1) = \begin{cases} P(H)P(2|H) & \text{for } s_1 = H \\ P(C)P(2|C) & \text{for } s_1 = C \end{cases} = \begin{cases} \mathbf{0.32} & \text{for } s_1 = H \\ 0.08 & \text{for } s_1 = C \end{cases}. \tag{10.20}$$

Since the maximum is achieved when $S_1 = H$, we have $\arg\max P(S_1|O_1) = H$. By using Eq. (10.20), we calculate μ as follows.

$$\begin{aligned}
\mu_2(S_2) = \max_{S_1} P(O_2|S_2)P(S_2|S_1)\mu_1(S_1) &= \begin{cases} P(1|H)P(H|H)\mu_1(H) \text{ for } s_1 = H, s_2 = H \\ P(1|H)P(H|C)\mu_1(C) \text{ for } s_1 = C, s_2 = H \\ P(1|C)P(C|H)\mu_1(H) \text{ for } s_1 = H, s_2 = C \\ P(1|C)P(C|C)\mu_1(C) \text{ for } s_1 = C, s_2 = C \end{cases} \\
&= \begin{cases} \mathbf{0.0384} \text{ for } s_1 = H, s_2 = H \\ 0.0064 \text{ for } s_1 = C, s_2 = H \\ \mathbf{0.0480} \text{ for } s_1 = H, s_2 = C \\ 0.0200 \text{ for } s_1 = C, s_2 = C \end{cases}.
\end{aligned} \tag{10.21}$$

Therefore, we obtain the most probable paths as $S_1 = H, S_2 = H$, and $S_1 = H, S_2 = C$. If we continue the iteration by the same way via Eq. (10.21),

$$\mu_3(S_3) = \max_{S_2} P(O_3|S_3)P(S_3|S_2)\mu_2(S_2) = \begin{cases} P(3|H)P(H|H)\mu_2(H) \text{ for } s_2 = H, s_3 = H \\ P(3|H)P(H|C)\mu_2(C) \text{ for } s_2 = C, s_3 = H \\ P(3|C)P(C|H)\mu_2(H) \text{ for } s_2 = H, s_3 = C \\ P(3|C)P(C|C)\mu_2(C) \text{ for } s_2 = C, s_3 = C \end{cases}$$

$$= \begin{cases} \mathbf{0.009216} & \text{for } s_1 = H, s_2 = H \\ 0.007680 & \text{for } s_1 = C, s_2 = H \\ 0.001152 & \text{for } s_1 = H, s_2 = C \\ \mathbf{0.002400} & \text{for } s_1 = C, s_2 = C \end{cases}. \tag{10.22}$$

Hence, we get the most probable paths as $S_1 = H, S_2 = H, S_3 = C$ and $S_1 = H, S_2 = H, S_3 = H$. Finally, by using the results in Eq. (10.22), we reach

$$\max_{S_3} \mu_3(S_3) = \max_{S_3} \{\mu_3(H), \mu_3(C)\} = \max\{0.009216, 0.002400\} = 0.009216. \tag{10.23}$$

As a result we conclude that the path, which presents $S_1 = H, S_2 = H, S_3 = H$, is the most probable path if the sequence of observations is $O = (2, 1, 3)$.

10.2.4 Baum–Welch Algorithm: Estimating the Model Parameters

The Baum–Welch forward backward method [5] is an iterative algorithm which is also a special case of the expectation–maximization approach. Here, we start the calculation with an initial guess of the parameters and by using data in hand, we aim to make better estimates for the model parameters λ iteratively until λ converges. In these computations, we use the following expression for the estimator of the transition probability between the ith and the jth variables, i.e., states.

$$\hat{a}_{ij} = \frac{\text{Expected number of transitions from } i \text{ to } j}{\text{Expected number of transitions from } i}. \tag{10.24}$$

To find these expectations, we apply the following equation.

$$\begin{aligned} \xi_t(i, j) &= P(S_t = i, S_{t+1} = j|O, \lambda), \\ &= \frac{P(S_t = i, S_{t+1} = j, O|\lambda)}{P(O|\lambda)}, \\ &= \frac{\alpha_t(S_t = s_i)a_{ij}b_{S_{t+1}=s_j, o_{t+1}}\beta_{t+1}(S_{t+1} = s_j)}{\sum_{j=1}^N \alpha_t(S_t = s_j)\beta_t(S_t = s_j)}. \end{aligned} \tag{10.25}$$

In Eq. (10.25), we can find the denominator by using only the forward or only the backward algorithm too. Then, by computing the function ξ, we can write the estimator for a_{ij} as

$$\hat{a}_{ij} = \frac{\sum_{t=1}^{T-1} \xi_t(i, j)}{\sum_{t=1}^{T-1} \sum_{k=1}^{N} \xi_t(i, k)}. \tag{10.26}$$

Similarly, to estimate the emission probability matrix B we can use,

$$\hat{b}_j(o_k) = \frac{\text{Expected number of times being in state } s_j \text{ and observing } o_k}{\text{Expected number of times being in state } s_j}. \tag{10.27}$$

Accordingly, the meaning of γ_t is

$$\gamma_t(j) = P(S_t = j | O, \lambda) = \frac{P(S_t = j, O|\lambda)}{P(O|\lambda)} = \frac{\alpha_t(S_t = s_j)\beta_t(S_t = s_j)}{\sum_{j=1}^{N} \alpha_t(S_t = s_j)\beta_t S_t = s_j}. \tag{10.28}$$

Finally, we can write our estimate for b_j as follows.

$$\hat{b}_j(o_k) = \frac{\sum_{t=1 \text{ st } O_k=o_k}^{T} \gamma_t(j)}{\sum_{t=1}^{T} \gamma_t(j)}. \tag{10.29}$$

Also, we can state the estimate of the initial probability π as

$$\hat{\pi}_i = \gamma_1(i). \tag{10.30}$$

10.3 Application

10.3.1 Data Description

In our work, we use the HIV-1 protease cleavage 746 dataset [27]. The data contain the lists of octamers (8 amino acids) and a flag depending on whether the HIV-1 protease will cleave in the central position (between amino acids 4 and 5). There are 401 cleaved and 345 non-cleaved octamers. We also use the physicochemical properties of amino acids from the AAIndex database [22]. In this database, there are 544 properties taken as continuous variables for each amino acid. We discard 14 of them since they contain null values.

10.3.2 Creation of States

In modeling our data via HMM, there are 8 bit observation sequences where each observation is from a set of 20 standard amino acids, namely, A, R, N, D, C, Q, E, G, H, I, L, K, M, F, P, S, T, W, Y, and V. Each observation has a hidden state behind it, which we form by using physicochemical properties of amino acids. Furthermore, we accept that if we replace an amino acid of a cleaved sequence with another amino acid having similar properties, it is more likely that the new sequence will also be a cleaved sequence. Therefore, we group amino acids according to the similarities based on physicochemical properties and use that information as their hidden states in the model.

After discarding features with null values, we have 530 features that can be taken for the analyses. On the other hand, when a clustering algorithm is used with a large number of features, typically, it can perform poorly due to outliers or highly correlated variables. Therefore, in our calculation, we implement some feature selection methods to decrease the number of features. For this purpose, initially, we group the features via the same AAIndex data by treating features as instances. Here, we use the k-means [9], k-medoids [26], and the hierarchical clustering [21] techniques. In the k-means and the hierarchical clustering approaches, we form a subset of features by choosing a variable randomly from each cluster. On the other side, in the k-medoids, we select the cluster medoids as the cluster representatives. Furthermore, we try different numbers of feature subsets that change from 30 to 60 with an increment of 5 in order to detect the optimal number of subsets. Additionally, we construct the model without performing any feature selection and compare it with the models with the feature selection in order to observe the effect of these clusterings in modeling.

Thereby, by using the underlying subsets of features, we group amino acids to create the states. Here, we accept that an amino acid can share many different properties with multiple groups. Thus, we prefer the fuzzy clustering [1], rather than classical clustering approaches, in our analyses. In this way, an amino acid can belong to more than one cluster with a membership degree between 0 and 1, and the sum of the membership degrees adding to 1. Among alternative fuzzy approaches, we select the most well-known ones, namely, the fuzzy k-means [1], Gustafson and Kessel-like fuzzy k-means [8], and the fuzzy k-medoids [17]. The fuzzy k-means is similar to usual the k-means method, where the Gustafson–Kessel-like fuzzy k-means considers non-spherical clusters too. In the fuzzy k-medoids, the medoids are being taken as the cluster representatives instead of artificial means. Finally, we assign amino acids to a state if the membership degree is greater than 0.1.

On the other hand, in our calculations, since there are 20 amino acids to cluster, the number of clusters cannot be more than 10. Whereas, we see that when the number of clusters is less than 5, too much information is lost. Therefore, we try and compare the number of states from 5 to 10 in order to detect the optimal number. Also, we standardize the features before clustering to avoid any bias caused by the

Table 10.1 States created by the fuzzy k-medoids approach with 9 number of states using 60 features which are determined via the hierarchical clustering

States	Amino acids
1	R, N, D, C, Q, M, P, W, Y
2	C, I, M, F, P, W, Y
3	R, N, D, C, G, P, S, T, Y
4	A, N, C, G, M, P, S
5	I, L, M, W
6	C, G, I, M, V
7	R, N, K
8	D, E, P
9	N, C, H, W, Y

variance of the features. In Table 10.1, we present 9 states created by the fuzzy k-medoids method using 60 features chosen by the hierarchical clustering.

10.3.3 Initialization of the EM Algorithm

On the other side, while doing the inference on emission, starting and transition probabilities, we use the Baum–Welch EM algorithm, which can converge to a local maximum instead of the global maximum [5]. Therefore, we give a clever starting point to the algorithm in order to increase the probability of reaching the global maximum in our calculation. Hence, we use data in hand to make a good prediction in the following way.

1. Calculation of initial probabilities: To calculate the starting probability of a state, we count all the sequences in the training data which start with amino acids that this state includes. Then, for all states, we divide them to the sum of these counts to turn these counts into probabilities. For example, let us say we have 15 sequences in the training data where 5 of them starting with A, 2 starting with S, and 8 starting with V. State 3 contains A, therefore, its count is counted by 2; State 4 contains both A and S, therefore, its count is set to $2 + 5$; and State 6 contains V, hence, its count equals to 8. Thus, the probability of the first state being State 3 is found as $2/17$, the first state being State 4 is equated to $7/17$ and the first state being State 6 is computed as $8/17$ while all other values in the vector Π being 0.
2. Calculation of emission probabilities: To estimate the probability of observing an amino acid given a state, we apply the following procedure. With the 0.9 probability, we observe one of the amino acids that this state includes, and with the 0.1 probability, other amino acids that this state does not include. As an example, State 1 includes 9 amino acids where the probability 0.9 is equally distributed among them, each having probability $0.9/9$, and the rest of the amino acids has the 0.1 probability equally distributed among them, each having the probability $0.1/11$.
3. Calculation of transition probabilities: We know the corresponding states for each amino acid of our sequences in the data. For example, Table 10.2 shows the

corresponding states for amino acids of the sequence AIMALKMR. For example, as seen in Table 10.2, there are 2 transitions from State 5 to State 5 and there is a 1 transition from State 5 to State 2 given only the sequence AIMALKMR. In this way, we count all transitions coming from all sequences in the training set. Afterwards, for each state, we sum all transitions from this state to all states including itself and divide counts of all transitions from this state to this number. Accordingly, we can turn it into a probability distribution. In case of the sum being 0, the probability of this row is taken equally distributed as $1/N$ for each state.

Figures 10.2 and 10.3 show the count matrix and the transition matrix produced by using only the sequence AIMALKMR.

10.3.4 Modeling the Data via HMM

In our calculation, we initially split cleaved and non-cleaved data into 90% of training and 10% of the test data. Then, we only use the test data after finding the optimal model parameters through training. Using initializations for the model parameters, we apply the Baum–Welch EM algorithm with the 1000 maximum numbers of iterations, and the convergence criteria for the change of the log-likelihood equal to 0.001. Furthermore, in all analyses, we conduct the R programming language and we utilize the `aphid` R package for the calculation.

Table 10.2 Corresponding states of the amino acids in the sequence AIMALKMR

Sequence	A	I	M	A	L	K	M	R
States	4	2	1	4	5	7	1	1
		5	2				2	3
		6	4				4	7
			5				5	
			6				6	

	State1	*State2*	*State3*	*State4*	*State5*	*State6*	*State7*	*State8*	*State9*
State1	1	0	1	1	0	0	1	0	0
State2	2	1	1	1	1	1	1	0	0
State3	0	0	0	0	0	0	0	0	0
State4	1	1	1	1	2	1	1	0	0
State5	2	1	1	2	1	1	2	0	0
State6	2	1	1	2	1	1	1	0	0
State7	1	1	0	1	1	1	0	0	0
State8	0	0	0	0	0	0	0	0	0
State9	0	0	0	0	0	0	0	0	0

Fig. 10.2 Counts of the transitions between states produced from the sequence AIMALKMR

	State1	*State2*	*State3*	*State4*	*State5*	*State6*	*State7*	*State8*	*State9*
State1	1/4	0	1/4	1/4	0	0	1/4	0	0
State2	2/8	1/8	1/8	1/8	1/8	1/8	1/8	0	0
State3	1/9	1/9	1/9	1/9	1/9	1/9	1/9	1/9	1/9
State4	1/8	1/8	1/8	1/8	2/8	1/8	1/8	0	0
State5	2/10	1/10	1/10	2/10	1/10	1/10	2/10	0	0
State6	2/9	1/9	1/9	2/9	1/9	1/9	1/9	0	0
State7	1/5	1/5	0	1/5	1/5	1/5	0	0	0
State8	1/9	1/9	1/9	1/9	1/9	1/9	1/9	1/9	1/9
State9	1/9	1/9	1/9	1/9	1/9	1/9	1/9	1/9	1/9

Fig. 10.3 Transition matrix produced from the sequence AIMALKMR

Moreover, the optimum values of the hyper-parameters are selected by using the 10-fold cross validation on the training data. The cross validation is used to reduce the bias stems from the random selection of data. Accordingly, the training data are divided into 10 folds and 9 of them are used for the training data as well as the last one is used for the validation data. Finally, we repeat this process 10 times until we utilize all 10 folds as the validation data.

To classify the sequence as cleaved or non-cleaved, two separate HMMs are trained on the cleaved and non-cleaved datasets, respectively. We declare these sequences as cleaved if the likelihood of belonging to cleaved HMM is greater than the non-cleaved HMM and vice versa. This way, we calculate the false positive (FP), false negative (FN), true positive (TP), and the true negative (TN) values. To measure the quality of our classification, we compute the precision(pre), recall (rec), accuracy (acc), Matthews correlation coefficient (MCC), and the F-measure (F). The formulas of these measures are also represented as below.

$$\text{Precision} = \frac{\text{TP}}{\text{TP+FP}},$$

$$\text{Recall} = \frac{\text{TP}}{\text{TP+FN}},$$

$$\text{Accuracy} = \frac{\text{TP+TN}}{\text{TP+TN+FP+FN}},$$

$$\text{MCC} = \frac{(\text{TP} \times \text{TN}) - (\text{FP} \times \text{FN})}{\sqrt{(\text{TP+FP}) \times (\text{TP+FN}) \times (\text{TN+FP}) \times (\text{TN+FN})}},$$

$$\text{F-measure} = 2 \times \frac{\text{Precision} \times \text{Recall}}{\text{Precision+Recall}}.$$

Lastly, after deciding the final model, to avoid over optimism caused by the overfitting, we declare results by using the test data whose final model has not been seen yet.

10.3.5 Results

In this section, for brevity, we refer to the fuzzy k-means method as fkm, Gustafson–Kessel-like fuzzy k-means as GKfkm and the fuzzy k-medoids approach as fkmed.

1. Effect of the number of states when other hyper-parameters are fixed: The number of states does not have a linear effect on the accuracy values when other parameters are fixed. Figure 10.4 shows the accuracy as a function of the number of states and the feature selection methods with the number of features used as 60 (Fig. 10.4a–c) or no features used (Fig. 10.4d). As seen in Fig. 10.4, when GKfkm is used, the accuracy increases with the number of states except from the case when the number of states changes from 5 to 6. In that case there is a slight decrease. On the other hand, there is no common pattern when other techniques are used. In our analyses, totally, we perform 3 feature selection techniques, 7 different number of features, and 3 state selection techniques, which makes a total of $3 \times 7 \times 3 = 63$ possible cases. This number becomes 66 when we include cases when we do not implement any feature selection. Out of these 66 cases, 40 of them give the best accuracy when the number of states is 10, 14 of them give the best accuracy when the number of states is 9, 5 of them give the optimal accuracy when the number of states is 8, followed by number of states 5, 6, and 7, respectively. Hence, we conclude that the large numbers of states produce more accurate results.
2. Effect of the number of features on the accuracy: Figs. 10.5, 10.6, and 10.7 show the accuracy as a function of the number of features for different feature selection and state selection methods. As seen in the figures, the effect of the number of features highly depends on the methods used. Moreover, the change in the number of features does not have an effect on the accuracy when GKfkm is applied. Also, fkm is very robust to the changes in the number of features only when the hierarchical feature selection method is performed. Finally, we observe that there is no common pattern for the other methods in the analyses.
3. Effect of the feature selection methods on the accuracy: As seen in Fig. 10.8, when the fkm state selection method is implemented, the k-medoids method gives the best results almost all the time except a few cases. Whereas, the hierarchical method gives poor results and this result does not change with the number of features. The k-means method, however, does not follow a common pattern, being worse than the k-medoids approach for most of times, but having higher accuracy for a few cases. As seen in Fig. 10.9, the feature selection methods are more robust and none of them is particularly better than each other when the fkmed method is applied. Additionally, the change in the feature selection method does not have an effect on the accuracy when GKfkm is used. Only exception is seen when the number of states is 5 in such a way that the accuracy values for this method do not change within our range for the number of features, but change when we do not implement any feature selection. Lastly, when the hierarchical feature selection method is used with the fkm state

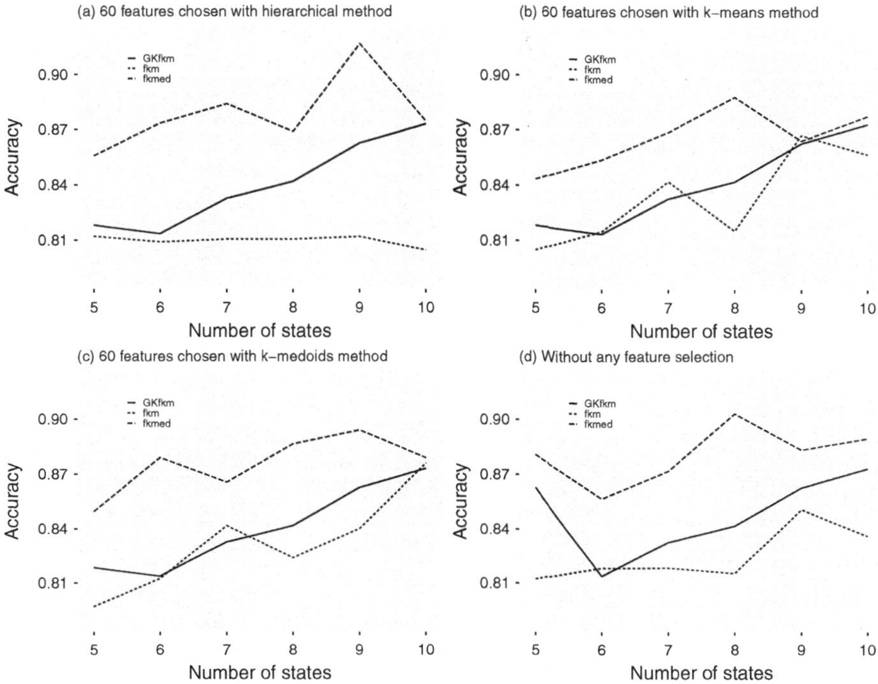

Fig. 10.4 Effect of the number of states on the accuracy values

selection, the number of features affects the accuracy slightly and the accuracy value is very poor.

4. Effect of the state selection methods on the accuracy: When we compare the state selection methods, we see that GKfkm is very robust to the number of features and the feature selection methods, but the accuracy changes when the number of states change. fkm is also very robust to the changes in the number of features when the hierarchical feature selection method is performed. As seen in Fig. 10.5, when the hierarchical feature selection method is applied, fkmed always gives the best results, GKfkm produces worse results, and fkm shows the worst outcomes. This is an expected finding since the fkmed method is more robust to outliers in the data and the GKfkm method captures non-spherical patterns unlike fkm. As seen in Fig. 10.6, a similar pattern appears when the k-means feature selection is implemented, except in some cases fkm surpasses fkmed and GKfkm. Moreover, fkm works more efficiently when it is used with the k-medoids feature selection method. As seen in Fig. 10.7, in some cases fkm gives better results than fkmed and on many cases it shows better results than GKfkm. Overall, there are 132 different cases when all other hyper-parameters are fixed except state selection methods. The fkmed method is the best among state selection methods 118 times out of 132 cases, followed by fkm which is the best 14 times, and GKfkm is never the best among other methods.

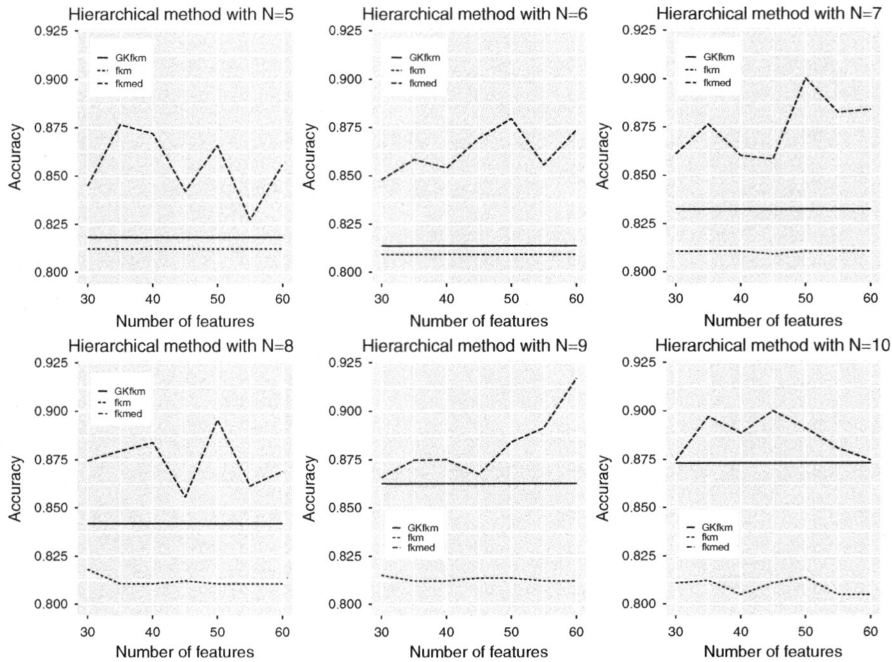

Fig. 10.5 Accuracy values for the hierarchical feature selection

5. Effect of the feature selection on the accuracy: When fkm is implemented for the state selection, only the k-medoids and sometimes the k-means feature selection give higher accuracy compared to no feature selection. When the GK-fkm state selection method is used, we observe that there is no difference between the feature selection and no feature selection findings. On the other hand, when fkmed is applied as the state selection, using all the dataset without any feature selection shows either the best or comparable results to the models with the feature selection approach.

As a results, at the end of the training process, we select the hierarchical feature selection method with 60 features and the fuzzy k-medoids state selection with 9 numbers of states as the optimal choices for our analyses. The associated states can be seen in Table 10.1. In the paper of Zhang et al. [37], a method, called the multiple property grouping, is suggested. We apply this method to our dataset and compare the results on both the 10-fold cross validation values on the training data and on the test data. The measures taken for the comparison are smaller on the test data than the training data since the model does not see the test data throughout the training process. As presented in Table 10.3, from the outcomes, it is observed that the proposed model gives better results than the multiple property grouping on both training and the test data on almost all measures except a slightly smaller value of the precision on the test data.

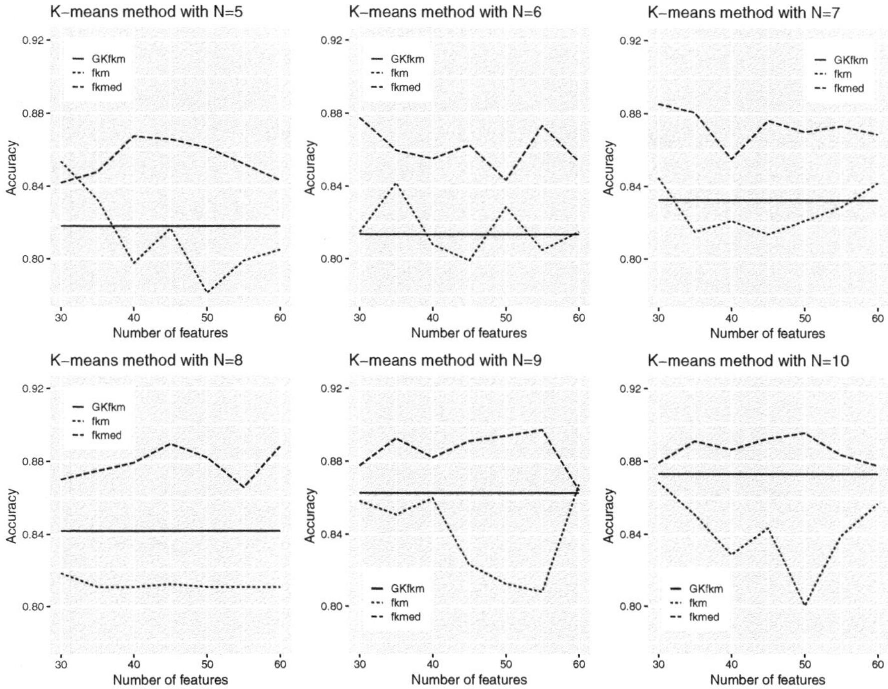

Fig. 10.6 Accuracy values for the k-means feature selection

Table 10.3 Comparison of the proposed model with the multiple property grouping the state selection method

Model	Precision	Recall	F-score	MCC	Accuracy
Proposed model training values	0.908	0.945	0.924	0.837	0.917
Multi property model training values	0.886	0.917	0.900	0.777	0.888
Proposed model test values	0.864	0.950	0.905	0.789	0.893
Multi property model test values	0.875	0.875	0.875	0.732	0.867

10.4 Conclusion

In this study, the hidden Markov model (HMM) has been used in order to detect the lock-and-key relationship in the Chip-seq data. In the application, we have initially explained the mathematical details of HMM in different stages of the estimation of the model parameters via the expectation–maximization method. Furthermore, we have investigated the effect of the clustering approaches in different aspects in the selection of the observations which are the sequence of amino acids and the states which are biophysical features of amino acids. In these analyses, we have conducted various methods from k-means and the hierarchical techniques to fuzzy methods. Among alternatives, we have chosen the fuzzy k-medoids methods due to

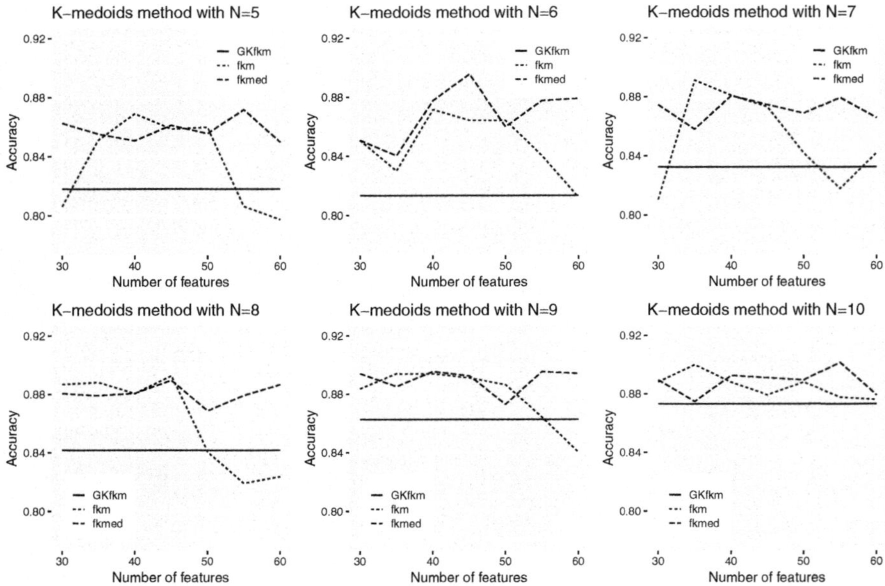

Fig. 10.7 Accuracy values for the k-medoids feature selection

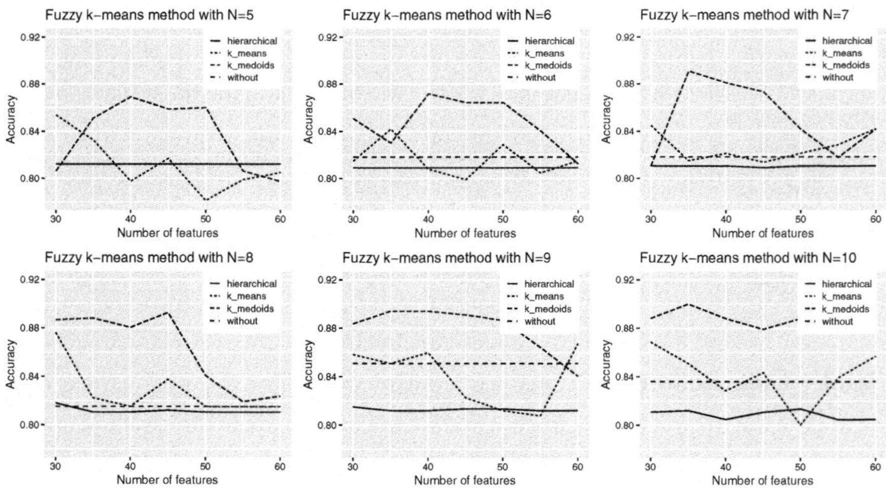

Fig. 10.8 Accuracy values for the fuzzy k-means state selection

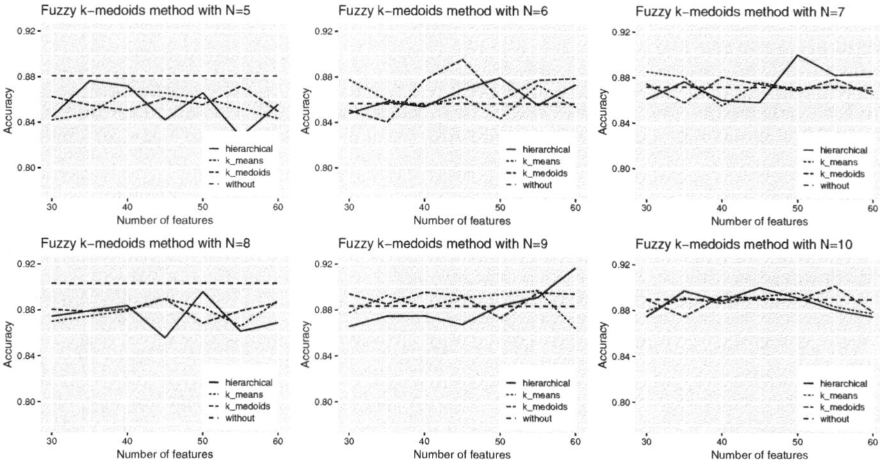

Fig. 10.9 Accuracy values for the fuzzy k-medoids state selection

the uncertain nature of the model construction and the higher accuracy of the results with respect to other approaches. Finally, we have evaluated the performance of all the suggested methods in different accuracy measures. From the findings, we have observed that HMM is promising to describe the selected benchmark Chiq-seq dataset, and the proposal clustering approaches have improved its accuracy based on accuracy of the estimates.

As the extension of this study in mathematical side, we consider to investigate the performance of the Bayesian inference in HMM and its more advance structure where the inference can be also applied via the variational approximation in order to diagnose AIDS in earlier stage. Because while the structure of Markov model becomes complex, the estimation of the model parameters cannot be solved via the sole frequentist approaches or simple iterative algorithms such as the Gibbs sampling although the complex model can more accurately capture the earlier levels of this disease and the effect of other risk factors causing AIDS. On the other side, as the extension of this study, we can consider the effect of AIDS in social side too. Because this illness has a significant impact on all levels of the society. The impact is particularly devastating not only for the individual who is infected, but also for the family and the wider community. The HIV/AIDS social work is changed fundamentally by the introduction of more effective medications which prolong the life of people living with this disease. The dominant themes in the related social work are transformed from loss and grief, to survival and living with the disease. Hereby, the social welfare policies mandate social workers to build a social capital in order to manage the impact of HIV and AIDS on communities. At this point, social workers may help diffuse HIV and AIDS information and showcase positive role-modeling behaviors, and provide members with material, emotional, and social supports [29, 32]. By combining these works with statistical modeling, it can be possible to clearly show the social effects and the factors of the AIDS disease on the patient simultaneously.

Acknowledgements The authors would like to thank the EU 7th Framework project, called COSTNET (Project no: CA15109), and the BAP project at the Middle East Technical University (Project no: BAP-08-11-2017-035) for their supports.

References

1. Bezdek J.: Pattern Recognition with Fuzzy Objective Function Algorithms. Plenum Press, New York (1981)
2. Cai, Y.D., Chou, K.C.: Artificial neural network model for predicting HIV protease cleavage sites in protein. Adv. Eng. Softw. **29**(2), 119–128 (1998)
3. Cai, Y., Liu, X., Xu, X., Chou, K.: Support vector machines for predicting HIV protease cleavage sites in protein. J. Comput. Chem. **23**, 267–274 (2002)
4. Chormungea, S., Jenab, S.: Correlation based feature selection with clustering for high dimensional data. J. Electr. Syst. Inf. Technol. (2018). doi: https://doi.org/10.1016/j.jesit.2017.06.004
5. Durbin, R., Eddy, S., Krogh, A., Mitchison, G.: Biological Sequence Analysis. Cambridge, UK (1998)
6. Eisner, J.: An interactive spreadsheet for teaching the forward-backward algorithm. In: Proc. of the ACL Workshop on Effective Tools and Methodologies for Teaching NLP and CL 10–18 (2002)
7. Gallo, R.C., Salahuddin, S.Z., Popovic, M., Shearer, G.M., Kaplan, M., Haynes, B.F., Palker, T.J., Redfield, R., Oleske, J., Safai, B., White, Cl., Foster, P., Markham, P.D.: Frequent detect on and isolation of cytopathic retroviruses (HTLV-III) from patients with AIDS and at risk for AIDS. Science **224**(4648), 500–503 (1984)
8. Gustafson, D.E., Kessel, W.C.: Fuzzy clustering with a fuzzy covariance matrix. Proc. IEEE CDC 761–766 (1978)
9. Hartigan, J.A., Wong, M.A.: Algorithm AS 136: A K-means clustering algorithm. J. R. Stat. Soc. Ser. C (Appl. Stat.) **28**, 100–108 (1979)
10. Jaeger, S., Chen, S.-S.: Information fusion for biological prediction. J. Data Sci. **8**, 269–288 (2010)
11. Jayavardhana Rama, G.L., Palaniswami, M.: Cleavage knowledge extraction in HIV-1 protease using hidden Markov model. In: Proc. 2nd International Conference on Intelligent Sensing and Information Processing, pp. 469–473 (2005)
12. Jianying, H., Brown, M.K., Turin, W.: HMM based online handwriting recognition. IEEE Trans. Pattern Anal. Mach. Intell. **18**, 1039–1045 (1996)
13. Juang, B., Rabiner, L.: Hidden Markov models for speech recognition. Technometrics **33**(3), 251–272 (1991)
14. Kim, G., Kim, Y., Lim, H., Kim, H.: An MLP-based feature subset selection for HIV-1 protease cleavage site analysis. Artif. Intell. Med. **48**, 83–89 (2010)
15. Kohl, N.E., Emini, E.A., Schlief, W.A., Davis, L.J., Heimbach, J., Dixon, R.A.F., Scolnik, E.M., Sigal, I.S.: Active human immunodeficiency virus protease is required for viral infectivity. Proc. Nutl. Sci. USA. **85**(15), 4686–4690 (1988)
16. Kouemou, G.L.: History and Theoretical Basics of Hidden Markov Models. Hidden Markov Models Przemyslaw Dymarski, IntechOpen (2011). doi: https://doi.org/10.5772/15205
17. Krishnapuram, R., Joshi, A., Nasraoui, O., Yi, L.: Low-complexity fuzzy relational clustering algorithms for Web mining. IEEE Trans. Fuzzy Syst. **9**(4), 595–607 (2001)

18. Miller, M., Schneider, J., Sathyanarayana, B.K., Toth, M.V., Marshall, G.R., Clawson, L., Selk, L., Kent, S.B.H., Wlodawer, A.: Structure of complex of synthetic HIV-l protease with a substrate-based inhibitor at 2.3 A resolution. Science **246**, 1149–1152 (1989)
19. Mitra, P., Murthy, C.A., Pal, S.K.: Unsupervised feature selection using feature similarity. IEEE Trans. Pattern Anal. Mach. Intell. **24**(3), 301–312 (2002)
20. Munch, K., Krogh, A.: Automatic generation of gene finders for eukaryotic species. BMC Bioinform. **7**, 263 (2006)
21. Murtagh, F.: Multidimensional Clustering Algorithms. Physica-Verlag (1985)
22. Nakai, K., Kidera, A., Kanehisa, M.: Cluster analysis of amino acid indices for prediction of protein structure and function. Protein Eng. **2**, 93–100 (1988)
23. Nanni, L.: Comparison among feature extraction methods for HIV-1 protease cleavage site prediction. Pattern Recognit. **39**(4), 711–713 (2006)
24. Niu, B., Yuan, X.C., Roeper, P., Su, Q., Peng, C.R., Yin, J.Y., Ding, J., Li, H., Lu, W.C.: HIV-1 protease cleavage site prediction based on two-stage feature selection method. Protein Pept. Lett. **20**, 290–298 (2013)
25. Pachter, L., Alexandersson, M., Cawley, S.: Applications of generalized pair hidden Markov models to alignment and gene finding problems. J. Comput. Biol. **9**, 389–399 (2002)
26. Park, H., Jun, C.: A simple and fast algorithm for k-medoids clustering. Expert Syst. Appl. **36**, 3336–3341 (2009)
27. Rögnvaldsson, T., You, L., Garwicz, D.: State of the art prediction of HIV-1 protease cleavage sites. Bioinformatics **31**, 1204–1210 (2015)
28. Schroff, R.W., Gottlieb, M.S., Prince, H.E., Chai, L.L., Fahey, J.L.: Immunological studies of homosexual men with immunodeficiency and Kaposi's sarcoma. Clin. Immunol. Immunopathol. **27**(3), 300–314 (1983)
29. Sesane, M., Geyer, S.: The perceptions of community members regarding the role of social workers in enhancing social capital in metropolitan areas to manage HIV and AIDS. Social Work **53**(1), 1–26 (2017)
30. Starner, T., Pentland, A.: Real-time American Sign Language recognition from video using hidden Markov models. In: Proc. of International Symposium on Computer Vision - ISCV, pp. 265–270 (1995)
31. Stultz, C.M.: Structural analysis based on state-space modeling. Protein Sci. **2**, 305–314 (1993)
32. Strug, D.L., Grube, B.A., Beckerman, N.L.: Challenges and changing roles in HIV/AIDS. Soc. Work Health Care **35**(4), 1–19 (2008)
33. Thompson, T.B., Chou, K.C., Zheng, C.: Neural network prediction of the HIV-1 protease cleavage sites. J. Theor. Biol. **177**(4), 369–379 (1995)
34. Turhal, U., Gök, M., Durgut, A.: Comparison among feature encoding techniques for HIV-1 protease cleavage specificity. Int. J. Intell. Syst. Appl. Eng. **3**(2), 62–66 (2015)
35. UNAIDS.: http://www.unaids.org/en/resources/fact-sheet
36. White, J.V.: Protein classification by stochastic modeling and optimal filtering of amino-acid sequences. Math. Biosci. **119**, 35–75 (1994)
37. Zhang, C., Bickis, M.G., Wu, F.X., Kusalik, A.J.: Optimally-connected hidden Markov models for predicting MHC-binding peptides. J. Bioinform. Comput. Biol. **4**(5), 959–980 (2006)

Chapter 11
A Numerical Approach for Variable Order Fractional Equations

Fatma Ayaz and İrem Bektaş Güner

11.1 Introduction

Since the last two or three decades, fractional calculus has become valuable tool in many branches of science and engineering. However, its history goes back to eighteenth century. Many scientists, including famous mathematicians such as Fourier (1822), Abel (1823–1826), Liouville (1822–1837), Riemann (1847), have contributed significant works for development of fractional calculus. There are many possible generalizations of $\frac{d^n f(x)}{dx^n}$, where n is not an integer, but the most important of these are the Riemann–Liouville and Caputo derivatives. The first of these appeared earlier than the others and was developed in works of Abel, Riemann, and Liouville in the first half of the nineteenth century. The mathematical theory of this derivative has been well established so far, but it has disadvantage that leads to difficulties especially for initial and boundary values, since in real world problems, these conditions cannot be described by fractional derivatives. Thus, the latter one, the Caputo derivative was derived by Caputo to eliminate the difficulties in identifying initial and boundary conditions. Both derivatives are very well known in the theory of fractional differential equations and the definitions of these derivatives will be given in the following section.

It has been proved that many physical processes can be well defined and modelled by fractional order differential equations. Moreover, fractional analysis provides many benefits for identifying and best modelling the physical systems which are suggested by scientists. Therefore, fractional order derivatives are much more suitable than the ordinary derivatives (see references [1–4]). For instance, it is not

F. Ayaz (✉) · İ. B. Güner
Faculty of Science, Gazi University, Ankara, Turkey
e-mail: fayaz@gazi.edu.tr

© Springer Nature Switzerland AG 2020

J. A. T. Machado et al. (eds.), *Numerical Solutions of Realistic Nonlinear Phenomena*, Nonlinear Systems and Complexity 31, https://doi.org/10.1007/978-3-030-37141-8_11

easy to explain abnormal diffusion behaviours by integer order differential equations since these processes appear abnormally with respect to time and space variables and it requires fractional models.

There are many application areas where these mathematical models are used and some of them can be listed here as physics, chemistry, biology, economics, control theory, signal and image processing, blood flow phenomenon, aerodynamics, fitting of experimental data, etc. Usually these models have complex nature; therefore, analytical solutions can only be obtained for certain classes of equations. Many numerical and approximate methods have been developed to solve these kinds of equations so far. Some of these methods are given as follows: finite difference approximation methods [5–10], fractional linear multistep methods [11–13], quadrature method [14–19], adomian decomposition method [20–22], variational iteration method [22, 23], differential transform method [24], Laplace perturbation method [25, 26], homotopy analysis method [27], etc. On the other hand, existing pure numerical techniques have usually first order convergency. However, it is well known that raising the order of convergency is a factor that increases the power of the method [28].

Nowadays, there are further developments in the analysis of fractional order differential equations and some studies are dealt with variable order fractional derivatives [29]. Thus, the need to develop more reliable methods in parallel with the developments in this field is inevitable.

The aim of this work is to use a second order convergent method to the variable fractional order multi-term differential equations similar to the work in [30] and to obtain reliable results. In the next section, second order convergent method will be mentioned and the theory of the method will be dealt with. Sections 11.3 and 11.4 are applications of the method for adding extra $y(t)$ and $y''(t)$ terms to the single-term equation. The last section is the conclusion.

11.2 Problem Definition and Integration Method for Variable Order Fractional Differential Equations

In this section, we first consider the following single-term initial value problem with fractional derivative, where $\alpha(t)$ is a function of time. Therefore, we can write the problem as

$$\begin{cases} {}_C D_{0,t}^{\alpha(t)} y(t) = f(t), \quad 0 \leq t \leq T, \\ \qquad y(0) = 0, \end{cases} \tag{11.1}$$

where $f(t)$ is a continuous function of t for a given interval. If $y(0) = \mu$, then by using the transformation $v(t) = y(t) - \mu$, we get $y(0) = 0$. In Eq. (11.1), $\alpha(t)$ denotes the order of variable fractional Caputo derivative, namely ${}_C D$, and this derivative is defined as,

$$
{}_{C}D_{0,t}^{\alpha(t)}y(t)=\frac{1}{\Gamma(1-\alpha(t))}\int_{0}^{t}(t-s)^{-\alpha(t)}y'(s)ds. \tag{11.2}
$$

We also recall the variable fractional order Riemann–Liouville derivative, ${}_{RL}D$ as

$$
{}_{RL}D_{0,t}^{\alpha(t)}y(t)=\frac{1}{\Gamma(1-\alpha(t))}\frac{d}{dt}\int_{0}^{t}(t-s)^{-\alpha(t)}y(s)ds. \tag{11.3}
$$

Consequently, by the following lemma, we see the relation between the Riemann–Liouville and the Caputo derivatives.

Lemma 11.1 *If $y(t)\in C[0,\infty)$ then, similar to the constant order fractional operators, the relation between variable order Caputo and Riemann–Liouville fractional derivatives is*

$$
{}_{C}D_{0,t}^{\alpha(t)}y(t)={}_{RL}D_{0,t}^{\alpha(t)}[y(t)-y(0)]. \tag{11.4}
$$

In Eq. (11.1), since the initial condition is $y(0)=0$, this follows that

$$
{}_{C}D_{0,t}^{\alpha(t)}y(t)={}_{RL}D_{0,t}^{\alpha(t)}y(t). \tag{11.5}
$$

Consequently, for convenience, the Caputo derivative is replaced by Riemann–Liouville derivative in Eq. (11.1). To obtain a numerical approach to the Riemann–Liouville variable order fractional derivative by a second order convergent method, we first call the shifted Grünwald approximation of a function $y(t)$

$$
\mathscr{A}_{\tau,p}^{\alpha(t)}y(t)=\frac{1}{\tau^{\alpha(t)}}\sum_{k=0}^{\infty}g_{k}^{\alpha(t)}y(t-(k-p)\tau), \tag{11.6}
$$

where, for $k\geq 0$,

$$
g_{k}^{\alpha(t)}=(-1)^{k}\binom{\alpha(t)}{k}.
$$

Now, the second order convergent method for Riemann–Liouville variable order derivative is defined as by the following theorem (see [30]).

Theorem 11.1 *Let $y(t)\in L^{1}(R)$ and its Riemann–Liouville derivative be ${}_{RL}D_{-\infty,t}^{\alpha(t)+2}y(t)$. For $\forall t_k\in R$, the Fourier transform of this derivative in $L^{1}(R)$ is [10]*

$$
\mathscr{D}_{\tau,p,q}^{\alpha(t)}y(t)=\frac{\alpha(t)-2q}{2(p-q)}\mathscr{A}_{\tau,p}^{\alpha(t)}y(t)+\frac{2p-\alpha(t)}{2(p-q)}\mathscr{A}_{\tau,q}^{\alpha(t)}y(t). \tag{11.7}
$$

Therefore,

$$\mathscr{D}^{\alpha(t_k)}_{\tau,p,q} y(t) = {}_{RL} D^{\alpha(t_k)}_{-\infty,t} y(t) + O(\tau^2), \tag{11.8}$$

where p and q are integers and $p \neq q$.

Proof From the definition of $\mathscr{A}^{\alpha(t)}_{\tau,p} y(t)$ as in Eq. (11.6), we write

$$\begin{aligned}\mathscr{D}^{\alpha(t_k)}_{\tau,p,q} y(t) &= \frac{\alpha(t_k) - 2q}{2(p-q)} \frac{1}{\tau^{\alpha(t_k)}} \sum_{k=0}^{\infty} g_k^{\alpha(t_k)} y(t-(k-p)\tau) \\ &+ \frac{2p - \alpha(t_k)}{2(p-q)} \frac{1}{\tau^{\alpha(t_k)}} \sum_{k=0}^{\infty} g_k^{\alpha(t_k)} y(t-(k-q)\tau).\end{aligned} \tag{11.9}$$

If the Fourier transform is applied to both sides of Eq. (11.9), the following expression is obtained

$$\begin{aligned}\mathscr{F}\left\{\mathscr{D}^{\alpha(t_k)}_{\tau,p,q} y(t); w\right\} &= \frac{1}{\tau^{\alpha(t_k)}} \sum_{k=0}^{\infty} g_k^{\alpha(t_k)} \left[\frac{\alpha(t_k) - 2q}{2(p-q)} e^{-iw(k-p)\tau}\right. \\ &\left.+ \frac{2p-\alpha(t_k)}{2(p-q)} e^{-iw(k-q)\tau}\right] \mathscr{F}(w) \\ &= \frac{1}{\tau^{\alpha(t_k)}} \left[\frac{\alpha(t_k) - 2q}{2(p-q)} (1-e^{-iw\tau})^{\alpha(t_k)} e^{iw\tau p}\right. \\ &\left.+ \frac{2p-\alpha(t_k)}{2(p-q)} (1-e^{-iw\tau})^{\alpha(t_k)} e^{iw\tau q}\right] \mathscr{F}(w) \\ &= (iw)^{\alpha(t_k)} \left[\frac{\alpha(t_k) - 2q}{2(p-q)} W_p(iw\tau) + \frac{2p-\alpha(t_k)}{2(p-q)} W_q(iw\tau)\right] \mathscr{F}(w),\end{aligned} \tag{11.10}$$

where $\mathscr{F}(w)$ is the Fourier transform of $y(t)$ and we writing

$$\begin{aligned}W_r(z) &= \left(\frac{1-e^{-z}}{z}\right)^{\alpha(t_k)} e^{rz} \\ &= 1 + \left(r - \frac{\alpha(t_k)}{2}\right) z + O(z^2), \qquad r = p, q,\end{aligned} \tag{11.11}$$

denoting,

$$\hat{g}\{w,\tau\} = \mathscr{F}\left\{\mathscr{D}^{\alpha(t_k)}_{\tau,p,q} y; w\right\} - \mathscr{F}\left\{{}_{RL} D^{\alpha(t_k)}_{-\infty,t} y; w\right\}$$

and by using Eqs.(11.10)–(11.11) and we have

$$\left|\mathscr{D}^{\alpha(t_k)}_{\tau,p,q}y(t) - {}_{RL}D^{\alpha(t_k)}_{-\infty,t}y(t)\right| = |g| \leq \frac{1}{2\pi}\int\limits_R |\hat{g}(w,\tau)|dw \leq C\|(iw)^{\alpha(t_k)+2}F(w)\|_{L^1}\tau^2$$
$$= O(\tau^2).$$

This completes the proof [30].

11.2.1 Numerical Method

To solve Eq. (11.1) numerically, we discretize the time domain, $t \in [0, T]$ by $\tau = \frac{T}{N}$, where N is an integer and $\alpha(t_k) = \alpha_k$ denotes the varying order fractional derivative with $t_k = k\tau, k = 0, 1, 2, 3 \ldots, N$. Moreover, choosing $(p, q) = (0, -1)$, then by using Eq. (11.9) we have

$$\frac{\alpha(t) - 2q}{2(p-q)} = \frac{2+\alpha(t)}{2}$$

and

$$\frac{2p - \alpha(t)}{2(p-q)} = -\frac{\alpha(t)}{2}.$$

Now, the second order convergent method can be given as follows [30]:

$$\begin{cases} \tau^{-\alpha_k}\sum_{j=0}^{k} w_j^{\alpha_k} y^{k-j} = f(t_k), \quad 1 \leq k \leq N \\ y^0 = 0 \end{cases}, \tag{11.12}$$

where, if $k = 0$, then $w_0^{\alpha_k} = (\frac{2+\alpha_k}{2})g_0^{\alpha_k}$,
otherwise, $w_j^{\alpha_k} = (\frac{2+\alpha_k}{2})g_j^{\alpha_k} - (\frac{\alpha_k}{2})g_{j-1}^{\alpha_k}$, $k \geq 1$ and $g_j^{\alpha_k} = (-1)^j \begin{pmatrix} \alpha_k \\ n \end{pmatrix}$.

11.2.2 Stability Criteria of the Method

This part deals with the stability of the method and the following lemma holds.

Lemma 11.2 *Being $\alpha_k \in (0, 1)$, then the coefficients, $w_j^{\alpha_k}$ in Eq. (11.12) satisfy the following properties:*

$$\begin{cases} w_0^{\alpha_k} = \frac{2+\alpha_k}{2}, \qquad w_j^{\alpha_k} < 0,\ j \geq 1 \\ \sum_{j=0}^{\infty} w_j^{\alpha_k} = 0,\ -\sum_{j=1}^{k} w_j^{\alpha_k} < w_0^{\alpha_k},\ k \geq 1. \end{cases} \tag{11.13}$$

Theorem 11.2 *Let $y(t) \in C[0, \infty)$ denotes exact and $\{y^k | k = 0, 1, 2, 3 \dots N\}$ numerical solution of Eq. (11.1) respectively, then the following inequality holds:*

$$|y^k| \leq \frac{5}{(1-\alpha_{\min})2^{\alpha_{\min}}} k^{\alpha_{\min}} \tau^{\alpha_{\min}} \max_{1 \leq m \leq k} |f(t_m)|. \tag{11.14}$$

Proof According to the Lemma 11.2, we know that

$$w_0^{\alpha_k} = \frac{2+\alpha_k}{2}, \, w_j^{\alpha_k} < 0, \, j \geq 1.$$

Hence, by arranging Eq. (11.12), we have

$$w_0^{\alpha_k} y^k = \sum_{j=1}^{k-1} (-w_j^{\alpha_k}) y^{k-j} + \tau^{\alpha_k} f(t_k), \qquad 1 \leq k \leq N. \tag{11.15}$$

For $k = 1$, we can write

$$|y^1| = |w_0^{\alpha_1}|^{-1} \tau^{\alpha_1} |f(t_1)| \leq \frac{5}{(1-\alpha_{\min})2^{\alpha_{\min}}} \tau^{\alpha_{\min}} |f(t_1)|.$$

Now, we have to show that Eq. (11.14) is also valid for $j = 1, 2, 3, \dots, k-1$. Hence, taking the absolute value of Eq. (11.15) and writing Eq. (11.14) into this inequality then we obtain

$$\begin{aligned}
w_0^{\alpha_k} |y^k| &\leq \left[\sum_{j=1}^{k-1} (-w_j^{\alpha_k}) |y^{k-j}| + \tau^{\alpha_k} |f(t_k)| \right] \\
&\leq \sum_{j=1}^{k-1} (-w_j^{\alpha_k}) \frac{5}{(1-(\alpha_{\min})2^{\alpha_{\min}}(k-j)^{\alpha_{\min}} \tau^{\alpha_{\min}}} \max_{1 \leq m \leq k-j} |f(t_m)| + \tau^{\alpha_{\min}} |f(t_k)| \\
&\leq \left[\sum_{j=1}^{k-1} (-w_j^{\alpha_k}) \frac{5}{(1-\alpha_{\min})2^{\alpha_{\min}}} k^{\alpha_{\min}} + 1 \right] \tau^{\alpha_{\min}} \max_{1 \leq m \leq k} |f(t_m)| \qquad (11.16) \\
&\leq \left\{ \left[w_0^{\alpha_k} - \frac{1-\alpha_{\min}}{5} \left(\frac{2^{\alpha_{\min}}}{k^{\alpha_{\min}}} \right) \right] \frac{5}{(1-\alpha_{\min})} 2^{\alpha_{\min}} + 1 \right\} \tau^{\alpha_{\min}} \max_{1 \leq m \leq k} |f(t_m)| \\
&= \frac{5 w_0^{\alpha_k}}{(1-\alpha_{\min})2^{\alpha_{\min}}} k^{\alpha_{\min}} \tau^{\alpha_{\min}} \max_{1 \leq m \leq k} |f(t_m)|.
\end{aligned}$$

Therefore, we get

$$|y^k| \leq \frac{5}{(1-\alpha_{\min})2^{\alpha_{\min}}} k^{\alpha_{\min}} \tau^{\alpha_{\min}} \max_{1 \leq m \leq k} |f(t_m)|.$$

As a result, by mathematical induction, Eq. (11.14) is valid for all $1 \leq k \leq N$ (See [30]).

Theorem 11.3 *Let $y(t) \in C[0,\infty)$ denote the exact solution and $\{y(t_k)|k = 0,1,2,3\ldots N\}$ define the values of y at t_k. Let us also denote the numerical solution of Eq. (11.1) by $\{y^k|k = 0,1,2,3\ldots N\}$ at particular points t_k. Therefore, absolute error in each step is denoted by $e^k = y(t_k) - y^k, k = 0,1,\ldots N$. Hence, the following relation holds:*

$$|e^k| \leq \frac{5c}{(1-\alpha_{\min})2^{\alpha_{\min}}} T^{\alpha_{\min}} \tau^2,$$

where c is a positive constant independent from τ.

Proof The proof of the theorem is given as in [30]. The error of Eq. (11.12) is

$$\begin{cases} \tau^{-\alpha_k} \sum_{j=0}^{k} w_j^{\alpha_k} e^{k-j} = R^k, \; 1 \leq k \leq N, \\ e^0 = 0. \end{cases} \tag{11.17}$$

This requires that $|R^k| \leq c\tau^2$. Then, by using Theorem 11.1 and Theorem 11.2, we write

$$\begin{aligned} |e^k| &\leq \frac{5c}{(1-\alpha_{\min})2^{\alpha_{\min}}} k^{\alpha_{\min}} \tau^{\alpha_{\min}} \max_{1\leq m\leq k} |R^m| \\ &\leq \frac{5c}{(1-\alpha_{\min})2^{\alpha_{\min}}} T^{\alpha_{\min}} \tau^2. \end{aligned}$$

This completes the proof.

11.2.3 Numerical Example

So far, a second order convergent method has been considered for approximating the Riemann– Liouville derivative, where the maximum error and the order of the convergency are obtained from the following formulas:

$$E_\infty(\tau) = \max_{0\leq k\leq N} |y(t_k) - y^k|,$$

$$order_\infty(\tau) = \log_2\left(\frac{E_\infty(2\tau)}{E_\infty(\tau)}\right).$$

To see the efficiency of the method the following example has been considered here. All numerical calculations have been done within MATLAB (R2015b).

Example 11.1 Assuming that $0 < \alpha(t) < 1$ and $T = 1$. Now, we can solve the following initial value problem [30]:

$$_{C}D_{0,t}^{\alpha(t)}y(t) = \frac{3t^{1-\alpha(t)}}{\Gamma(2-\alpha(t))} + \frac{2t^{2-\alpha(t)}}{\Gamma(3-\alpha(t))}, \quad 0 \leq t \leq T \tag{11.18}$$

$$y(0) = 0. \tag{11.19}$$

The exact solution of the problem is known as $y(t) = 3t + t^2$ and two different values of $\alpha(t)$ will be considered here:

Case 1: $\alpha(t) = \frac{1}{2}t$,
Case 2: $\alpha(t) = sin(t)$.

Consequently, by using the following numerical scheme:

$$\tau^{-\alpha_k} \sum_{j=0}^{k} w_j^{\alpha_k} y^{k-j} = f(t_k), 1 \leq k \leq N,$$

and taking $y^0 = 0$, numerical results are obtained. These results have been shown by tables. Tables 11.1 and 11.3 show the difference between exact and numerical solutions of the problem for $\tau = \frac{1}{16}$ and $\tau = \frac{1}{32}$. In Table 11.1, the first case $\alpha(t) = \frac{1}{2}t$ has been used. Moreover, in Table 11.3, the second case, $\alpha(t) = sin(t)$ was applied. Tables 11.2 and 11.4 denote the maximum error and order of convergency results for $\alpha(t) = \frac{1}{2}$ and $\alpha(t) = sin(t)$, respectively. Figure 11.1 shows both numerical and exact solutions in the same plot. It is clear that analytical and numerical solutions overlap.

11.3 Multi-term Variable Order Fractional Equations

In this section we will apply the second order convergent method to a new class of variable fractional order differential equations. With additional terms, we will have multi-term variable fractional order differential equation. First, we will apply $y(t)$ term to Eq. (11.1). Hence, the following initial value problem, Eq. (11.20), will be considered here:

$$\begin{cases} _{C}D_{0,t}^{\alpha(t)}y(t) + ay(t) = f(t), 0 \leq t \leq T \\ y(0) = 0. \end{cases} \tag{11.20}$$

For convenience, taking $a = 1$ and each t_k is in the discretized time domain, the following numerical scheme holds:

Table 11.1 The difference between the numerical and exact values of Example 11.1 for $T = 1$ and $\alpha(t) = \frac{1}{2}t$. The calculations have been performed for both $N = 16$, $N = 32$

$\tau = \frac{1}{16}$		$\tau = \frac{1}{32}$	
Numerical value	Exact value	Numerical value	Exact value
0.190969333717	0.191406250000	0.094615423638	0.094726562500
0.390168425750	0.390625000000	0.191289608582	0.191406250000
0.597204270428	0.597656250000	0.289922357608	0.290039062500
0.812056502729	0.812500000000	0.390509189828	0.390625000000
1.034722151963	1.035156250000	0.493049429619	0.493164062500
1.265200437577	1.265625000000	0.597542896551	0.597656250000
1.503491075175	1.503906250000	0.703989525702	0.704101562500
1.749593937717	1.750000000000	0.812389288867	0.812500000000
2.003508960064	2.003906250000	0.922742172037	0.922851562500
2.265236105578	2.265625000000	1.035048167506	1.035156250000
2.534775352514	2.535156250000	1.149307270671	1.149414062500
2.812126687793	2.812500000000	1.265519478597	1.265625000000
3.097290103864	3.097656250000	1.383684789304	1.383789062500
3.390265596989	3.390625000000	1.503803201392	1.503906250000
3.691053166230	3.691406250000	1.625874713838	1.625976562500
3.999652812814	4.000000000000	1.749899325873	1.750000000000
		1.875877036904	1.875976562500
		2.003807846471	2.003906250000
		2.133691754215	2.133789062500
		2.265528759855	2.265625000000
		2.399318863176	2.399414062500
		2.535062064016	2.535156250000
		2.672758362260	2.672851562500
		2.812407757834	2,812500000000
		2.954010250700	2.954101562500
		3.097565840850	3.097656250000
		3.243074528307	3.243164062500
		3.390536313121	3.390625000000
		3.539951195366	3.540039062500
		3.691319175139	3.691406250000
		3.844640252556	3.844726562500
		3.999914427756	4.000000000000

$$\begin{cases} \tau^{-\alpha_k} \sum_{j=0}^{k} w_j^{\alpha_k} y^{k-j} = f(t_k) - y(t_k), \quad 1 \le k \le N, \\ y^0 = 0. \end{cases} \tag{11.21}$$

Therefore, the following theorem is valid.

Theorem 11.4 *Let $y(t) \in C[0, \infty)$ denote the exact solution and $\{y(t_k)|k = 0, 1, 2, 3 \dots N\}$ define the values of y at t_k. Let us also denote the numerical*

Table 11.2 Maximum error and order of convergency results for different values of τ in Example 11.1, where $T = 1$ and $\alpha(t) = \frac{1}{2}t$

τ	$E_\infty(\tau)$	$order_\infty(\tau)$
$\frac{1}{4}$	$6.585728e-03$	
$\frac{1}{8}$	$1.754860e-03$	1.9079
$\frac{1}{16}$	$4.565743e-04$	1.9424
$\frac{1}{32}$	$1.167049e-04$	1.9679
$\frac{1}{64}$	$2.967417e-05$	1.9755

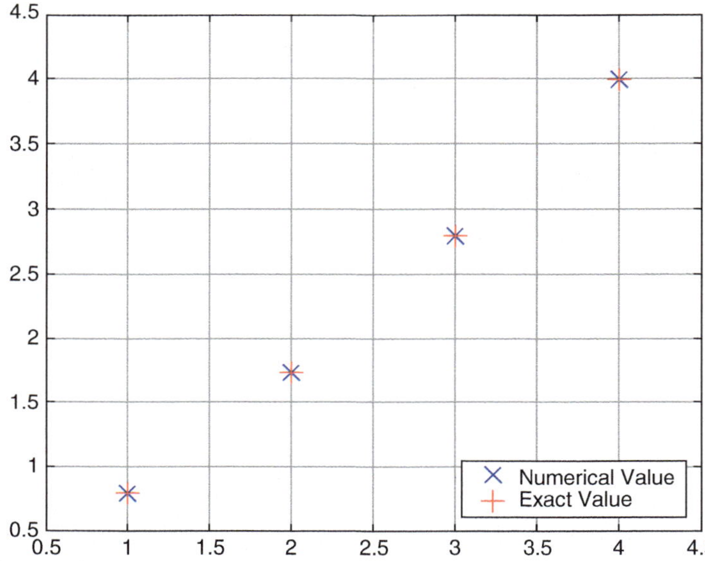

Fig. 11.1 Comparison of the numerical and exact solutions of $y(t)$ of Example 11.1, where $T = 1$, $\alpha(t) = \frac{1}{2}t$

solution of Eq. (11.1) by $\{y^k | k = 0, 1, 2, 3 \ldots N\}$ *at particular points* t_k. *Therefore, maximum error in each step is denoted by* $e^k = y(t_k) - y^k, k = 0, 1, \ldots N$. *Hence, the following relation holds:*

$$|e^k| \leq \frac{5c}{(1-\alpha_{\min})2^{\alpha_{\min}}} T^{\alpha_{\min}} \tau^2,$$

where c is a positive constant independent from τ.

Proof By using Eq. (11.12), the proof of this theorem can be performed easily same as the proof of Theorem 11.3.

Following example denotes that the second order convergent method is still valid for multi-term variable fractional order differential equations.

Table 11.3 Comparison of the numerical and exact solutions of $y(t)$ at particular point t^k in Example 11.1, where $T = 1$, $\alpha(t) = sin(t)$. The calculations have been performed for both $N = 16$, $N = 32$

$\tau = \frac{1}{16}$		$\tau = \frac{1}{32}$	
Numerical value	Exact value	Numerical value	Exact value
0.190458811746	0.191406250000	0.094495229084	0.094726562500
0.389552027574	0.390625000000	0.191152917929	0.191406250000
0.596522633108	0.597656250000	0.289776299993	0.290039062500
0.811321611175	0.812500000000	0.390355236098	0.390625000000
1.033941750135	1.035156250000	0,492888269766	0.493164062500
1.264381028701	1.265625000000	0.597375029086	0.597656250000
1.502638440292	1.503906250000	0.703815383037	0.704101562500
1.748713235941	1.750000000000	0.812209273587	0.812500000000
2.002604749102	2.003906250000	0.922556668864	0.922851562500
2.264312355507	2,265625000000	1.034857547135	1.035156250000
2.533835469305	2.535156250000	1.149111890582	1.149414062500
2.811173548221	2.812500000000	1.265319682707	1.265625000000
3.096326099295	3.097656250000	1.383480907242	1.383789062500
3.389292682744	3.390625000000	1.503595547725	1.503906250000
3.690072913517	3.691406250000	1.625663587402	1.625976562500
3.998666460891	4.000000000000	1.749685009261	1.750000000000
		1.875659796152	1.875976562500
		2.003587930921	2.003906250000
		2.133469396549	2.133789062500
		2.265304176291	2.265625000000
		2.399092253795	2.399414062500
		2.534833613201	2.535156250000
		2.672528239233	2.672851562500
		2.812176117265	2.812500000000
		2.953777233375	2.954101562500
		3.097331574386	3.097656250000
		3.242839127892	3.243164062500
		3.390299882272	3.390625000000
		3.539713826696	3.540039062500
		3.691080951124	3.691406250000
		3.844401246297	3.844726562500
		3.999674703724	4.000000000000

Table 11.4 Absolute errors and order of convergency for different values of τ in Example 11.1, where $T = 1$ and $\alpha(t) = sin(t)$

τ	$E_\infty(\tau)$	$order_\infty(\tau)$
$\frac{1}{4}$	$2.470109e-02$	
$\frac{1}{8}$	$5.602080e-03$	2.1513
$\frac{1}{16}$	$1.333539e-03$	2.0707
$\frac{1}{32}$	$3.253162e-04$	2.0353
$\frac{1}{64}$	$8.032494e-05$	2.0179

Table 11.5 Maximum error and order of convergency results for different values of τ in Example 11.2, where $T = 1$ and $\alpha(t) = \frac{1}{2}$

τ	$E_\infty(\tau)$	$order_\infty(\tau)$
$\frac{1}{4}$	$9.201325e-03$	
$\frac{1}{8}$	$2.623652e-03$	1.81026
$\frac{1}{16}$	$7.261612e-04$	1.85321
$\frac{1}{32}$	$1.960721e-04$	1.88890
$\frac{1}{64}$	$5.190739e-05$	1.91737

Example 11.2 In Eq. (11.20), assuming that $f(t) = t^2 + \frac{2}{\Gamma(\frac{5}{2})}t^{\frac{3}{2}}$ and $T = 1$, then, the exact solution of the problem is known as $y(t) = t^2$. But this solution is known for only $\alpha(t) = \frac{1}{2}$. To compare numerical results with exact ones, only $\alpha = \frac{1}{2}$ case has been considered here. Numerical calculations have been performed for different values of τ, and a code is written in MATLAB (R2015b). The following table, Table 11.5, lists maximum error and the order of convergency for different values of τ and Table 11.6 compares the numerical and exact solutions for $\tau = \frac{1}{16}$ and $\tau = \frac{1}{32}$.

11.4 Addition of $y''(t)$ Term to Variable Order Fractional Differential Equations

In this section, by using the second order convergent method which is given by Eq. (11.12), we will develop a hybrid method for wider classes of differential equations. By adding $y(t)$ and $y''(t)$ terms to Eq. (11.1), then we will have multi-term fractional differential equations. To solve the multi-term fractional equation, we approximate the second order derivative with the central differences and the fractional derivative term is evaluated as it is given in Eq. (11.12). Therefore, we will consider the following initial value problem:

Table 11.6 The difference between numerical and exact values of Example 11.1 for $T = 1$ and $\alpha(t) = sin(t)$. The calculations have been performed for both $N = 16$, $N = 32$

$\tau = \frac{1}{16}$		$\tau = \frac{1}{32}$	
Numerical value	Exact value	Numerical value	Exact value
0.004569025	0.003906250	0.001150760	0.000976562
0.016351161	0.015625000	0.004102322	0.003906250
0.035851178	0.035156250	0.008980809	0.008789062
0.063157151	0.062500000	0.015809364	0.015625000
0.098281108	0.097656250	0.024591676	0.024414062
0.141222933	0.140625000	0.035328044	0.035156250
0.191981309	0.191406250	0.048018296	0.047851562
0.250555215	0.250000000	0.062662261	0.062500000
0.316943955	0.316406250	0.079259815	0.079101562
0.391147053	0.390625000	0.097810869	0.097656250
0.473164167	0.472656250	0.118315358	0.118164062
0.562995042	0.562500000	0.140773233	0.140625000
0.660639483	0,660156250	0,165184459	0.165039062
0.766097340	0.765625000	0.191549003	0.191406250
0.879368490	0.878906250	0.219866843	0.219726562
1.000452834	1.000000000	0.250137959	0.250000000
		0.282362334	0.282226562
		0.316539954	0.316406250
		0.352670807	0.352539062
		0.390754883	0.390625000
		0.430792174	0.430664062
		0.472782672	0.472656250
		0.516726370	0.516601562
		0.562623262	0.562500000
		0.610473343	0.610351562
		0.660276608	0.660156250
		0.712033054	0.711914062
		0.765742676	0.765625000
		0.821405470	0.821289062
		0.879021435	0.878906250
		0.938590567	0.938476562
		1.000112863	1.000000000

$$\begin{cases} {}_C D_{0,t}^{\alpha(t)} y(t) = f(t) + a(t)y''(t) + b(t)y(t), \quad 0 \le t \le T, \\ y(0) = 0, \\ y'(0) = 0. \end{cases} \tag{*}$$

Therefore, at particular values of t_k, Eq. (*) is written as

$$\tau^{-\alpha_k} \sum_{j=0}^{k} w_j^{\alpha_k} y^{k-j} = f(t_k) + ay''(t_k) + by(t_k). \tag{11.22}$$

Assuming that a and b are constants and for simplicity, we take their values as $1, -1$ respectively. Now recalling central finite difference approximation to the second order derivative:

$$f''(t_k) = \frac{f(t_{k+1}) - 2f(t_k) + f(t_{k-1})}{h^2}, \tag{11.23}$$

and substituting this into Eq. (11.22), then the last form of the numerical scheme is obtained easily. The method will be applied to following example (see[19]).

Example 11.3 Consider the differential equation in Eq. (*) as follows:

$$D^{\alpha(t)} y(t_k) = y''(t_k) - y(t_k) + f(t_k). \tag{11.24}$$

Since the exact results are known for only $\alpha = \frac{1}{2}$, for comparing the numerical results with exact ones, we will also use $\alpha(t) = \frac{1}{2}$ in the calculations. Substituting the finite difference approximation to the second order ordinary derivative, then we have

$$\tau^{-\alpha_k} \sum_{j=0}^{k} w_j^{\alpha_k} y^{k-j} = \frac{y(t_{k+1}) - 2y(t_k) + y(t_{k-1})}{h^2} - y(t_k) + f(t_k). \tag{11.25}$$

The exact solution of the problem is known as $y(t) = t^2$, when

$$f(t) = t^2 - 2 + \frac{2}{\Gamma(\frac{5}{2})} t^{\frac{3}{2}}. \tag{11.26}$$

Hence,

$$\tau^{-\alpha_k} \sum_{j=0}^{k} w_j^{\alpha_k} y^{k-j} = \frac{y(t_{k+1}) - 2y(t_k) + y(t_{k-1})}{h^2} - y(t_k) + {t_k}^2 - 2 + \frac{2}{\Gamma(\frac{5}{2})} {t_k}^{\frac{3}{2}}, \tag{11.27}$$

where $h = \tau$. As a result, arranging Eq. (11.27) again, we obtain

$$y(t_{k+1}) = h^2\tau^{-\alpha_k}\sum_{j=0}^{k} w_j^{\alpha_k} y^{k-j} + 2y(t_k) - y(t_{k-1})$$
$$+h^2 y(t_k) - h^2[t_k{}^2 - 2 + \frac{2}{\Gamma(\frac{5}{2})} t_k{}^{\frac{3}{2}}]. \quad (11.28)$$

Table 11.7 Results for Example 11.3, where $T = 1$ and $\alpha(t) = \frac{1}{2}$. Comparison of the numerical and exact values of $y(t)$ for two different step size, $N = 10$ and $N = 100$, respectively

t_k	$y(t_k) = t_k{}^2$	$N = 10$ için y^k	$N = 100$ için y^k
0.1	0.01	0.010000000	0.010000470
0.2	0.04	0,040000000	0.040000729
0.3	0.09	0.090235468	0.090000693
0.4	0.16	0.160461209	0.160000405
0.5	0.25	0.250691879	0.249999884
0.6	0.36	0.360931289	0.359999132
0.7	0.49	0.491183812	0.489998146
0.8	0.64	0.641454438	0.639996910
0.9	0.81	0.811748730	0.809995404
1.0	1.0	1.002072855	0.999993598
	e_∞	0.002072855	0.000006401

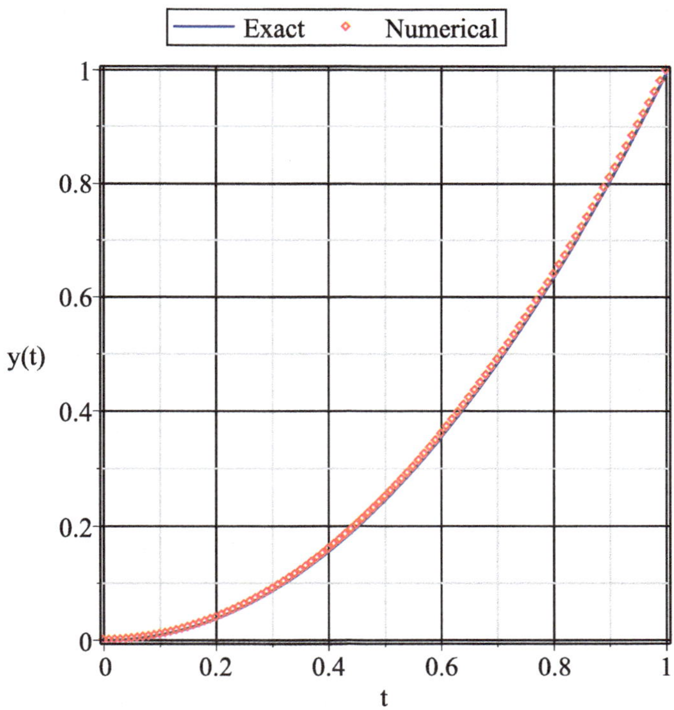

Fig. 11.2 For $\alpha(t) = \frac{1}{2}$ and $N = 100$. Comparison of numerical and exact values of $y(t)$ in Example 11.3

Table 11.7, shows the exact and numerical values of $y(t)$ at particular points of t for different step sizes where initial conditions are taken as in Eq. (*). Figure 11.2 illustrates that both exact and numerical results are in good agreement.

11.5 Conclusion

Here, we aimed to solve fractional-variable order differential equations and multi-term fractional order differential equations. We have used the second order convergent method as in [30]. The method is quite well when it is compared with the analytical solutions. For finer mesh, one can obtain more reliable results. As a result, the method can be applied to wider classes of fractional equations, variable order fractional differential equations.

References

1. Kilbas, A.A., Sirvastava, H.M., Trujillo, J.J.: Theory and Application of Fractional Differential Equations, vol. 204. North-Holland Mathematics Studies, Amsterdam (2006)
2. Ross, B.: Fractional Calculus and Its Applications, Lecture Notes in Mathematics, vol. 457. Springer (1975)
3. Oldham, K.B., Spanier, J.: The fractional calculus theory and applications of differentiation and integration of arbitrary order. Dower, New York (2006)
4. Cansiz, M.: Kesirli Diferensiyel Denklemler ve Uygulaması, Yüksek Lisans Tezi, Ege Üniversitesi Fen Bilimleri Enstitüsü, İzmir, pp. 16–33 (2010)
5. Podlubny, I.: Fractional Differential Equations. Academic Press, New York (1999)
6. Ciesielski, M., Leszcynski, J.: Numerical Simulations of Anomalous Diffusion. Computer Methods Mech Conference, Gliwice Wisla Poland (2003)
7. Yuste, S.B.: Weighted average finite difference methods for fractional diffusion equations. J. Comput. Phys. **1**, 264–274 (2006)
8. Odibat, Z.M.: Approximations of fractional integrals and Caputo derivatives. Appl. Math. Comput., 527–533 (2006)
9. Odibat, Z.M.: Computational algorithms for computing the fractional derivatives of functions. Math. Comput. Simul. **79**(7), 2013–2020 (2009)
10. Wang, Z., Vong, S.: Compact difference schemes for the modified anomalous fractional sub-diffusion equation and the fractional diffusion-wave equation. J. Comput. Phys. **277**, 1–15 (2014)
11. Ford, N.J., Joseph Connolly, A.: Systems-based decomposition schemes for the approximate solution of multi-term fractional differential equations. J. Comput. Appl. Math. **229**, 382–391 (2009)
12. Ford, N.J., Simpson, A.C.: The numerical solution of fractional differential equations: speed versus accuracy. Numer. Alg. **26**, 333–346 (2001)
13. Sweilam, N.H., Khader, M.M., Al-Bar, R.F.: Numerical studies for a multi-order fractional differential equations. Phys. Lett. A **371**, 26–33 (2007)
14. Diethelm, K.: An algorithm for the numerical solution of differential equations of fractional order. Electron. Trans. Numer. Anal. **5**, 1–6 (1997)
15. Diethelm, K., Walz, G.: Numerical solution of fractional order differential equations by extrapolation. J. Numer. Algorithms **16**, 231–253 (1997)

16. Diethelm, K., Ford, N.: Analysis of fractional differential equations. J. Math. Anal. Appl. **265**, 229–248 (2002)
17. Diethelm, K.: Generalized compound quadrature formulae for finite-part integrals. IMA J. Numer. Anal., 479–493 (1997)
18. Diethelm, K.: The Analysis of Fractional Differential Equations, pp. 85–185. Springer Pub., Germany (2004)
19. Görgülü, O.: Kesirli Mertebeden Diferensiyel Denklemler için Sayısal ve Yaklaşık Yöntemler, Yüksek Lisans Tezi, Gazi Üniversitesi Fen Bilimleri Enstitüsü, Ankara, pp. 13–43 (2017)
20. Momani, S., Odibat, Z.: Analytical solution of a time-fractional Navier-Stokes equation by Adomian decomposition method. App. Math. Comput. **177**, 488–494 (2006)
21. Odibat, Z., Momani, S.: Numerical approach to differential equations of fractional order. J. Comput. Appl. Math. **207**(1), 96–110 (2007)
22. Odibat, Z., Momani, S.: Numerical methods for nonlinear partial differential equations of fractional order. Appl. Math. Model. **32**, 28–39 (2008)
23. Odibat, Z., Momani, S.: Application of variational iteration method to equation of fractional order. Int. J. Nonlinear Sci. Numer. Simul. **7**, 271–279 (2006)
24. Erturk, V.S., Momani, S., Odibat, Z.: Application of generalized differential transform method to multi-order fractional differential equations. Commun. Nonlinear Sci. Numer. Simul. **13**(8), 1642–1654 (2008)
25. Yavuz, M., Ozdemir, N.: New Numerical Techniques for Solving Fractional Partial Differential Equations in Conformable Sense, Non-Integer Order Calculus and its Applications. Book Series: Lecture Notes in Electrical Engineering, vol. 496, pp. 49–62 (2019)
26. Yavuz, M., Ozdemir, N., Baskonus, M.H.: Solutions of partial differential equations using the fractional operator involving Mittag-Leffler kernel. Eur. Phy. J. Plus **133**(6), 215 (2018)
27. Yavuz, M., Ozdemir, N.: European vanilla option pricing model of fractional order without singular kernel. Fractal Fractional **2**(1), 3 (2018)
28. Er, F.N.: Kısmi Türevli Kesirli Mertebeden Lineer Schrodinger Denklemlerinin Sayısal Çözümleri, Doktora Tezi, İstanbul Külür Üniversitesi Fen Bilimleri Enstitüsü, İstanbul, pp. 14–63 (2015)
29. Weilbeer, M.: Efficient Numerical Methods for Fractional Differential Equations and Their Analytical Background, Doktora Tezi, Technische Universität Braunschweig, Germany, pp. 35–63 (2005)
30. Cao, J., Qiu, Y.: High order numerical scheme for variable order fractional ordinary differential equations. Appl. Math. Lett. **61**, 88–94 (2016)

Chapter 12
Evolution of Plane Curves via Lie Symmetry Analysis in the Galilean Plane

Zühal Küçükarslan Yüzbaşı, Ebru Cavlak Aslan, Dumitru Baleanu, and Mustafa Inc

12.1 Introduction

The symmetry analysis is one of the most important and efficient methods that can be used to analyze nonlinear differential equations. The theory of symmetry analysis has a big importance in geometry, mechanics, and physics. Moreover, there are several studies about Lie symmetry analysis performed for several equations, [3, 8, 10–12, 20]. Moreover, the problem of evolving curves in the plane is a quite interesting topic since it can be arranged different subjects on the same theoretical basis. One of them is a geometrical interpretation of integrable systems. There have been deep connections between the differential geometry of curve motions and the integrable systems have been examined in different space [4–6, 16, 18, 19]. Particularly, evolving curves problem has been studied via Lie group analysis or different aspects [1, 2, 17].

In this work, our aim is to study a general equation of the evolving curves by flow in the Galilean plane from the point of the symmetry analysis and get exact solutions of the equation which we obtain from the evolution of plane curve by the flow.

This manuscript is organized as follows: In Sect. 12.2, the evolution of plane curve equations by an inelastic and an elastic flow is determined in $\mathbf{G}_2$. In Sect. 12.3,

Z. K. Yüzbaşı · E. C. Aslan (✉) · M. Inc
Department of Mathematics, Firat University, Elazig, Turkey
e-mail: zuhal2387@yahoo.com.tr; ebrucavlak@hotmail.com; minc@firat.edu.tr

D. Baleanu
Department of Mathematics & Computer Science, Çankaya University, Ankara, Turkey

Institute of Space Sciences, Magurele, Bucharest, Romania
e-mail: dumitru@cankaya.edu.tr

© Springer Nature Switzerland AG 2020

J. A. T. Machado et al. (eds.), *Numerical Solutions of Realistic Nonlinear Phenomena*, Nonlinear Systems and Complexity 31, https://doi.org/10.1007/978-3-030-37141-8_12

we get the linear and nonlinear partial differential equations. Then the new classes of symmetry reductions for these equations are designated and exact solutions are obtained. Our results show that the symmetry analysis is a very efficient method to get the solution of these type equations.

12.2 Evolution of Plane Curves by Flow in G_2

We consider R^2 with the bilinear form

$$\langle \boldsymbol{\lambda}, \boldsymbol{\mu} \rangle = \begin{cases} \lambda_1\mu_1, & \text{if } \lambda_1 \neq 0 \text{ or } \mu_1 \neq 0 \\ \lambda_2\mu_2, & \text{if } \lambda_1 = 0 \text{ and } \mu_1 = 0 \end{cases}, \tag{12.1}$$

where $\boldsymbol{\lambda} = (\lambda_1, \lambda_2)$ and $\boldsymbol{\mu} = (\mu_1, \mu_2)$. Then we have the Galilean plane G_2. This is one of the three Cayley–Klein plane geometries with a parabolic measure of distance. Denote R^2 with the bilinear form (12.1), [14, 15].

Let $r : I \subseteq R \to G_2$ be a curve in Galilean plane given by $r = (s, x(s))$, where s is the arc length on r. Then, we have

$$T(s) = (1, x'(s)),$$
$$N(s) = \frac{1}{\kappa(s)}(0, x''(s)) = (0, 1),$$

where $\kappa(s) = x''(s)$.

Then the following moving bihedron of r is written as:

$$T'(s) = \kappa(s)N(s), \tag{12.2}$$
$$N'(s) = 0,$$

where T and N is said to be the tangent and principal normal of r in G_2 , [15].

Let us consider a smooth curve in Galilean plane. Suppose that u is the parameter along r in G_2. Let $r(u, t)$ represent the position vector of a point on r at t. The metric on r is

$$g(u, t) = \left\langle \frac{\partial r}{\partial u}, \frac{\partial r}{\partial u} \right\rangle. \tag{12.3}$$

The arc length along the curve is defined as follows:

$$s(u, t) = \int^u \sqrt{g(\sigma, t)}d\sigma, \quad \frac{\partial}{\partial s} = \frac{1}{\sqrt{g}}\frac{\partial}{\partial u}. \tag{12.4}$$

The motion of a point on the curve in G_2 is defined by

$$\frac{\partial r}{\partial t} = f_1 T + f_2 N, \tag{12.5}$$

where f_1 and f_2 are the velocities along the frame T and N, respectively. Also The motion is called local if $\{f_1, f_2\}$ depends only on local values of k and their derivatives [6].

Lemma 12.1 *The expression of the evolution equation for g is written as*

$$\dot{g} = 2g\frac{\partial f_1}{\partial s}. \tag{12.6}$$

Proof By differentiating (12.3) and (12.5) with respect to t and s, respectively, and since $\frac{\partial}{\partial u}, \frac{\partial}{\partial t}$ commute, then we get

$$\begin{aligned}
\dot{g} &= \frac{\partial g}{\partial t} = 2\left\langle \frac{\partial r}{\partial u}, \frac{\partial}{\partial t}\frac{\partial r}{\partial u}\right\rangle = 2g\left\langle \frac{\partial r}{\partial s}, \frac{\partial}{\partial s}\frac{\partial r}{\partial t}\right\rangle \\
&= 2g\left\langle T, \frac{\partial f_1}{\partial s}T + f_1 k N + \frac{\partial f_2}{\partial s}N\right\rangle \\
\dot{g} &= 2g\frac{\partial f_1}{\partial s}.
\end{aligned}$$

Lemma 12.2 *The expression of the evolution of the length of the curve has*

$$\frac{\partial s}{\partial t} = \int^{u} \frac{\partial f_1}{\partial s} d\sigma. \tag{12.7}$$

Proof From (12.4), we get

$$\frac{\partial s}{\partial t} = \int^{u} \frac{\dot{g}}{2\sqrt{g}} d\sigma. \tag{12.8}$$

Substituting (12.6) into (12.8), then the lemma holds.

Definition 12.1 A curve is said to be inelastic curve if its length is preserved, i.e.,

$$\frac{\partial s}{\partial t} = 0 \Leftrightarrow \dot{g} = 0. \tag{12.9}$$

Theorem 12.1 *The flow of r is inelastic if and only if f_1 is a constant.*

Proof $\Rightarrow$Suppose that the curve flow is inelastic. From (12.8), we get

$$\frac{\partial s}{\partial t} = \int^{u} \frac{\dot{g}}{2\sqrt{g}} d\sigma = 0. \tag{12.10}$$

Substituting (12.6) into (12.10), we have

$$\frac{\partial f_1}{\partial s} = 0,$$

this means that f_1 is a constant.

⇐Assume that f_1 is a constant, then from (12.6), it is easily shown that $\dot{g} = 0$, then $\frac{\partial s}{\partial t} = 0$. Thus the curve is inelastic.

Theorem 12.2 *For the curve flow* $\frac{\partial r}{\partial t} = f_1 T + f_2 N$. *If* $r(u,t)$ *be an elastic curve, then we get*

$$k_t = f_1 k_s + f_{2ss}. \tag{12.11}$$

Proof Let $r(u,t)$ be an elastic curve, that is, $\dot{g} \neq 0$. By differentiating (12.5) with respect to u, then we obtain

$$\begin{aligned} r_{tu} &= \sqrt{g} r_{ts} = \sqrt{g}\frac{\partial}{\partial s}\Big(\frac{\partial r}{\partial t}\Big) \\ &= \sqrt{g}\Big(\frac{\partial f_1}{\partial s} T + \Big(f_1 k + \frac{\partial f_2}{\partial s}\Big) N\Big). \end{aligned} \tag{12.12}$$

Since $r_u = \sqrt{g} r_s = \sqrt{g} T$, by differentiating with respect to t, then we get

$$r_{ut} = \sqrt{g}\Big(\frac{g_t}{2\sqrt{g}} T + T_t\Big). \tag{12.13}$$

From the condition of

$$r_{tu} = r_{ut},$$

we get

$$\frac{g_t}{2\sqrt{g}} = \frac{\partial f_1}{\partial s}, \tag{12.14}$$

$$T_t = (f_1 k + \frac{\partial f_2}{\partial s}) N. \tag{12.15}$$

By differentiating (12.15) with respect to u, we have

$$T_{tu} = \sqrt{g}\Big(\frac{\partial f_1}{\partial s} k + f_1 \frac{\partial k}{\partial s} + \frac{\partial^2 f_2}{\partial s^2}\Big) N. \tag{12.16}$$

By differentiating $T_u = \sqrt{g}T_s = \sqrt{g}(kN)$ with respect to t, we obtain

$$T_{ut} = \sqrt{g}\Big(\Big(\frac{g_t}{2\sqrt{g}}k + k_t\Big)N + kN_t\Big)\Big). \tag{12.17}$$

From the condition of

$$T_{tu} = T_{ut}$$

we obtain

$$k_t = f_1 k_s + f_{2ss}.$$

Corollary 12.1 *The flow of* r *is an inelastic then we also get*

$$k_t = f_1 k_s + f_{2ss},$$

such that f_1 *is a constant.*

12.3 Application via Lie symmetry Analysis

In this section, our starting point is to give Lie symmetry analysis for (12.15) with respect to an inelastic and an elastic flow for special choosing f_1 and f_2. Many researchers explain how to consider Lie symmetry analysis in many books, [7, 13]. We will now take into account the one parameter group of point transformations of the form

$$\begin{aligned}\widetilde{s} &\rightarrow s + \varepsilon\xi(s,t,k)\\ \widetilde{t} &\rightarrow t + \varepsilon\eta(s,t,k)\\ \widetilde{k} &\rightarrow u + \varepsilon\zeta(s,t,k),\end{aligned}$$

where ε is a group parameter. The vector field associated with the above group of transformations can be given by

$$V = \xi(s,t,k)\frac{\partial}{\partial s} + \eta(s,t,k)\frac{\partial}{\partial t} + \zeta(s,t,k)\frac{\partial}{\partial k}.$$

Case 12.3.1 For the inelastic flow of curve by taking $f_1 = 1$ and $f_2 = k_s$, (12.11) becomes

$$k_t = k_s + k_{sss}. \tag{12.18}$$

We can show it considering a well-established procedure that (12.18) admits the following infinitesimals:

$$\xi(s,t,k) = 2C_2t - C_2s + C_4,$$
$$\eta(s,t,k) = -3C_2t + C_3,$$
$$\zeta(s,t,k) = F(s,t) + C_1k,$$

where C_1, C_2, C_3, and C_4 are arbitrary constants and $F(s,t)$ is an arbitrary function of s and t. The algebra of Lie point symmetries of the above equation is generated by five vector fields

$$\begin{aligned} V_1 &= k\frac{\partial}{\partial k}, \\ V_2 &= -3t\frac{\partial}{\partial t} + (2t-s)\frac{\partial}{\partial s}, \\ V_3 &= \frac{\partial}{\partial t}, \\ V_4 &= \frac{\partial}{\partial s}, \\ V_5 &= F(s,t)\frac{\partial}{\partial k}. \end{aligned} \tag{12.19}$$

One may obtain that V_1, V_2, V_3, V_4, V_5 are closed under the Lie bracket.

In this equation, the infinitesimal generator is $V_1 + V_4 = k\frac{\partial}{\partial k} + \frac{\partial}{\partial s}$. From this generator, we have $k = e^s f(\rho),\ \rho = t$. So, (12.18) leads to reduced equation

$$f' = 2f \tag{12.20}$$

which constructs the solution of (12.20) as $f = e^{2t}C$, where C is an arbitrary constant.

Case 12.3.2 For the inelastic flow of curve by taking $f_1 = 1$ and $f_2 = \frac{k^2}{2}$, (12.11) becomes

$$k_t = k_s + k_s^2 + kk_{ss}. \tag{12.21}$$

We can show it considering a well-established procedure that (12.21) admits the following infinitesimals:

$$\xi(s,t,k) = C_1t + C_1s + C_2t + C_4,$$
$$\eta(s,t,k) = -C_2t + C_3,$$
$$\zeta(s,t,k) = (2C_1 + C_2)k,$$

where C_1, C_2, C_3, and C_4 are arbitrary constants.

The algebra of Lie point symmetries of the above equation is generated by four vector fields

$$V_1 = 2k\frac{\partial}{\partial k} + (t+s)\frac{\partial}{\partial s},$$

$$V_2 = -t\frac{\partial}{\partial t} + t\frac{\partial}{\partial s} + k\frac{\partial}{\partial k}, \tag{12.22}$$

$$V_3 = \frac{\partial}{\partial t},$$

$$V_4 = \frac{\partial}{\partial s}. \tag{12.23}$$

One may obtain that V_1, V_2, V_3, V_4 are closed under the Lie bracket. Then we can give the commutation of Lie algebra as follows:

$[.,.]$	V_1	V_2	V_3	V_4
V_1	0	0	$-V_4$	$-V_4$
V_2	0	0	$V_3 - V_4$	0
V_3	V_4	$-V_3 + V_4$	0	0
V_4	V_4	0	0	0

For the generator $V_3 + c_1 V_4 = \frac{\partial}{\partial t} + c_1\frac{\partial}{\partial s}$ (c_1 an arbitrary constant), we have $k = f(\rho),\ \rho = s - c_1 t$. If we substitute this function into (12.21), we reduce the following nonlinear ODE:

$$ff'' + f'^2 + f'(1 + c_1) = 0 \tag{12.24}$$

which creates the solution of (12.24) as $f = (-1-c_1)\rho + C$, where C is an arbitrary constant.

Case 12.3.3 For the elastic flow of curve by taking $f_1 = k^2$ and $f_2 = k_s$, (12.11) becomes

$$k_t = k^2 k_s + k_{sss}. \tag{12.25}$$

We can show it considering a well-established procedure that (12.25) admits the following infinitesimals:

$$\xi(s,t,k) = -C_1 s + C_3,$$

$$\eta(s,t,k) = -3C_1 t + C_2,$$

$$\zeta(s,t,k) = C_1 k,$$

where C_1, C_2 and C_3 are arbitrary constants. The algebra of Lie point symmetries of the above equation is generated by three vector fields

$$\begin{aligned} V_1 &= -3t\frac{\partial}{\partial t} + k\frac{\partial}{\partial k} - s\frac{\partial}{\partial s}, \\ V_2 &= \frac{\partial}{\partial t}, \\ V_3 &= \frac{\partial}{\partial s}. \end{aligned} \tag{12.26}$$

One may obtain that V_1, V_2, V_3 are closed under the Lie bracket. Then we can give the commutation of Lie algebra as follows:

$[.,.]$	V_1	V_2	V_3
V_1	0	$2V_2$	V_3
V_2	$-2V_2$	0	0
V_3	$-V_3$	0	0

For the generator $V_2+c_2V_3 = \frac{\partial}{\partial t}+c_2\frac{\partial}{\partial s}$ (c_2 an arbitrary constant), the invariants are obtained as $k = f(\rho)$ and $\eta = s - c_2t$. Equation (12.25) leads to reduced equation

$$f''' + f^2f' + c_2f' = 0. \tag{12.27}$$

Riccati–Bernoulli ODE method will be applied for (12.27).

Suppose that the solution of (12.27) is the solution of Riccati– Bernoulli equation

$$f' = bf + af^{2-m} + cf^m, \tag{12.28}$$

where a, b, and c are the constants [9]. Then the second and third derivatives of (12.28) yield

$$f'' = f^{-1-2m}\left(af^2 + cf^{2m} + bf^{1+m}\right)\left(a(2-m)f^2 + cmf^{2m} + bf^{1+m}\right). \tag{12.29}$$

$$\begin{aligned} f''' &= f^{-2-3m}(af^2 + cf^{2m} + bf^{m+1})(a^2(2m^2 - 7m + 6)f^4 + c^2m(2m-1)f^{4m} \\ &+ ab(m^2 - 5m + 6)f^{3+m} + (b^2 + 2ac)f^{2+2m} + bcm(m+1)f^{1+3m}). \end{aligned} \tag{12.30}$$

Substituting (12.29) and (12.30) into (12.27), we have

$$f^{-2-3m}(af^2 + cf^{2m} + bf^{m+1})(a^2(m-2)(2m-3)f^4 + c^2m(2m-1)f^{4m}$$

$$+ab(m-3)(m-2)f^{3+m} + bcm(1+m)f^{1+3m} + f^{2+2m}(b^2+2ac+c_2+f^2)). \tag{12.31}$$

Setting $m = 0$, the above equation is reduced to

$$(c + f(b+af))(b^2+2ac+c_2+f(6ab+f+6a^2f)) = 0. \tag{12.32}$$

Each coefficient of $f^i (i = 0, 1, 2, 3, 4)$ should be equal to zero. We obtain $a = -\frac{c_2}{2c}$, $b = 0$, and $c = i\sqrt{\frac{3}{2}}c_2$.

When $m \neq 1, a \neq 0$, and $b^2 - 4ac > 0$, the solution is

$$f(\eta) = \left(-\frac{b}{2a} - \frac{\sqrt{b^2-4ac}}{2a}\tanh\left[\frac{(1-m)\sqrt{b^2-4ac}}{2}(\rho + C)\right]\right)^{\frac{1}{1-m}}. \tag{12.33}$$

So, the soliton solution is

$$f(\eta) = \left(i\sqrt{3c_2}\tanh\left[\sqrt{\frac{c_2}{2}}(\rho + C)\right]\right). \tag{12.34}$$

Case 12.3.4 For the elastic flow of curve by taking $f_1 = k_{ss}$ and $f_2 = kk_s$, (12.11) becomes

$$k_t = 4k_sk_{ss} + kk_{sss}. \tag{12.35}$$

We can show it considering a well-established procedure that (12.35) admits the following infinitesimals:

$$\xi(s,t,k) = C_2s + C_4,$$
$$\eta(s,t,k) = C_1t + C_3,$$
$$\zeta(s,t,k) = -k(C_1 - 3C_2),$$

where C_1, C_2, and C_3 are arbitrary constants. The algebra of Lie point symmetries of the above equation is generated by four vector fields

$$\begin{aligned} V_1 &= t\frac{\partial}{\partial t} - k\frac{\partial}{\partial k}, \\ V_2 &= 3k\frac{\partial}{\partial k} + s\frac{\partial}{\partial s}, \\ V_3 &= \frac{\partial}{\partial t} \\ V_4 &= \frac{\partial}{\partial s}. \end{aligned} \tag{12.36}$$

One may obtain that V_1, V_2, V_3, V_4 are closed under the Lie bracket. Then we can give the commutation of Lie algebra as follows:

$[.,.]$	V_1	V_2	V_3	V_4
V_1	0	0	$-V_3$	0
V_2	0	0	0	$-V_4$
V_3	V_3	0	0	0
V_4	0	V_4	0	0

Firstly, for the generator $V_1 = t\frac{\partial}{\partial t} - k\frac{\partial}{\partial k}$, we have $k = f(\rho),\ \rho = s$. If we substitute this function into (12.35), we reduce the following nonlinear ODE:

$$1 + 4f'' + f''' = 0 \tag{12.37}$$

which creates the solution of (12.37) as $f = -\frac{s^2}{8} + \frac{1}{16}e^{-4s}C_1 + sC_2 + C_3$, where C_1, C_2 and C_3 are arbitrary constants.

Secondly, for the generator $V_2 = 3k\frac{\partial}{\partial k} + s\frac{\partial}{\partial s}$, we have $k = s^3 f(\rho),\ \rho = t$. If we substitute this function into (12.35), we reduce the following nonlinear ODE

$$f' - 78f^2 = 0 \tag{12.38}$$

which creates the solution of (12.38) as $f = \frac{1}{-C - 78s}$, where C is an arbitrary constant.

Finally, for the generator $V_3 + c_3 V_4 = \frac{\partial}{\partial t} + c_3\frac{\partial}{\partial s}$, we have $k = f(\rho),\ \rho = c_3 t - s$. If we substitute this function into (12.35), we reduce the following nonlinear ODE

$$c_3 f' + 4f'f'' + ff''' = 0. \tag{12.39}$$

In a similar way to Case 12.3.3, Riccati–Bernoulli ODE method will be applied for (12.39). So, the solution is

$$f(\rho) = \left(\frac{1}{a(m-1)(\rho + C)} - \frac{a}{b}\right)^{\frac{1}{1-m}}. \tag{12.40}$$

When $m = 2$, $a \neq 0$, and $b^2 - 4ac = 0$, the solution is

$$f(\rho) = (a(\rho + C)). \tag{12.41}$$

12.4 Conclusion

In this paper, we have studied the evolution of plane curve equations in the context of an inelastic and an elastic flow is determined in G_2. Under the different choices in these equations, we obtained the linear and nonlinear partial differential equations. Some of the equations acquired are solved via the Lie symmetry analysis. Besides, if we make an appropriate choice, then we get the Burger's equation from the evolution of plane curve equation. We also illustrated Fig. 12.1 in cases of $c_2 = 0.1$, $C = 1$ and Fig. 12.2 in cases of $c_2 = 1$, $C = 0.1$ for an evolution of plane curve equation in Case 12.3.3 (Figs. 12.3 and 12.4).

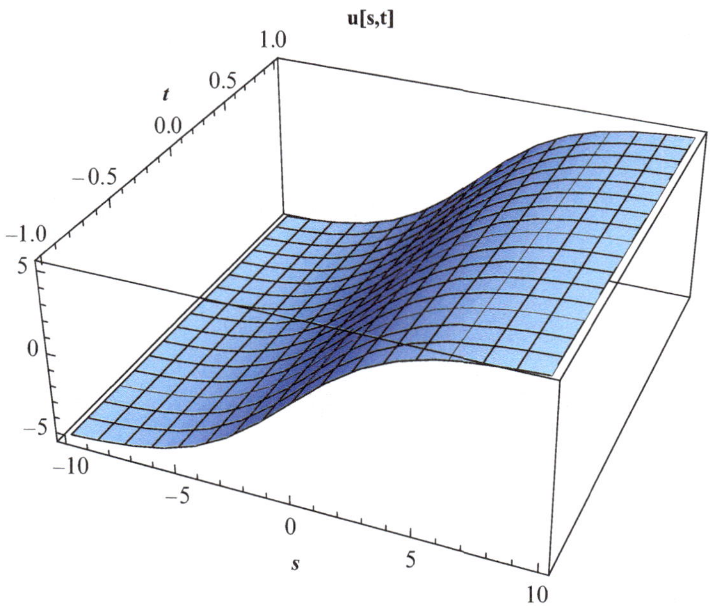

Fig. 12.1 The curvature of the curve for $c_2 = 0.1$, $C = 1$, $s \in [-10, 10]$, and $t \in [-1, 1]$ in 3D

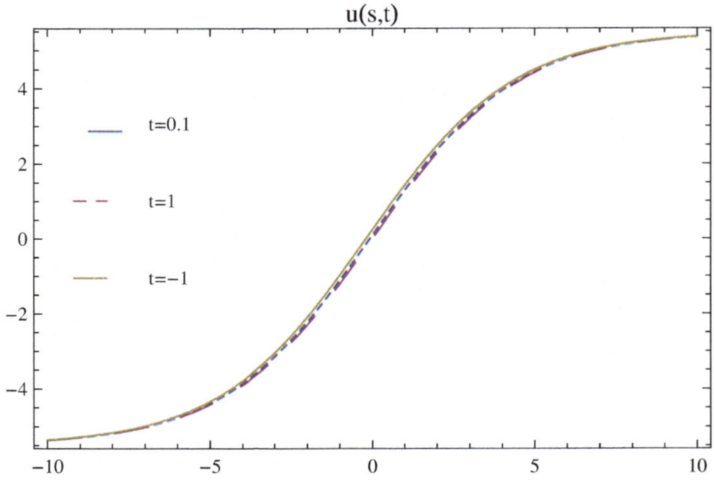

Fig. 12.2 The curvature of the curve for $c_2 = 0.1$, $C = 1$, $s \in [-10, 10]$, and $t \in [-1, 1]$ in 2D

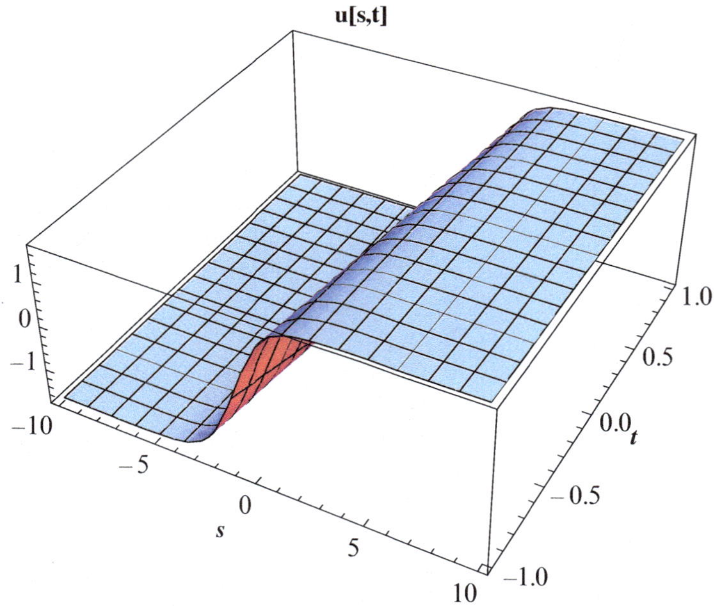

Fig. 12.3 The curvature of the curve for $c_2 = 1$, $C = 0.1$, $s \in [-10, 10]$, and $t \in [-1, 1]$ in 3D

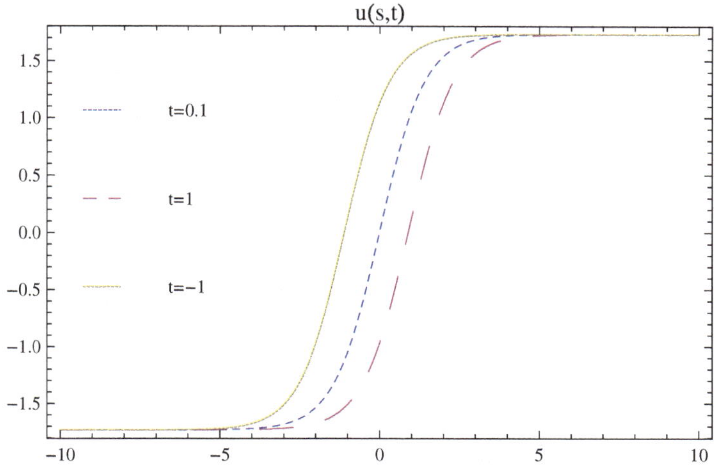

Fig. 12.4 The curvature of the curve for $c_2 = 1, C = 0.1, s \in [-10, 10]$, and $t \in [-1, 1]$ in 2D

References

1. Abdel-All, N.H., Abdel-Razek, M.A., Abdel-Aziz, H.S., Khalil, A.A.: Geometry of evolving plane curves problem via Lie group analysis. Stud. Math. Sci. **2**(1), 51–62 (2011)
2. Abdel-All, N.H., Mohamed, S.G., Al-Dossary, M.T.: Evolution of generalized space curve as a function of its local geometry. Appl. Math. **5**(15), 2381 (2014)
3. Abd-el-Malek, M.B., Amin, A.M.: Lie group method for solving the generalized Burgers, Burgers KdV and KdV equations with time-dependent variable coefficients. Symmetry **7**, 1816–1830 (2015)
4. Alkan, K., Anco, S.C.: Integrable systems from inelastic curve flows in 2 and 3-dimensional Minkowski space. J. Nonlinear Math. Phys. **23**(2), 256–299 (2016)
5. Akar, M., Yuce, S., Kuruoglu, N.: One parameter planar motion on the Galilean plane. Int. Electron. J. Geom. **6**(1), 79–88 (2013)
6. Akbiyik, M., Yüce, S.: The areas of the trajectory surface under the Galilean motions in the Galilean space. Int. J. Geo. Meth. Modern Phys. **15**(09), 1850162 (2018)
7. Bluman, G., Anco, S.: Symmetry and Integration Methods for Differential Equations, vol. 154. Springer Science Business Media, New York (2008)
8. Huang, D., Li, X., Yu, S.: Lie symmetry classification of the generalized nonlinear beam equation. Symmetry **9**(115) (2017). doi: https://doi.org/10.3390/sym9070115
9. Inc, M., Aliyu, A.I., Yusuf, A.: Solitons and conservation laws to the resonance nonlinear Shrödinger's equation with both spatio-temporal and inter-modal dispersions. Optik **142**, 509–522 (2017)
10. Janocha, D.D., Wacławczyk, M., Oberlack, M.: Lie symmetry analysis of the hopf functional-differential equation. Symmetry **7**, 1536–1566 (2015)
11. Liu, H., Li, J., Zhang, Q.: Lie symmetry analysis and exact explicit solutions for general Burgers' equation. J. Comput. Appl. Math. **228**(1), 1–9 (2009)
12. Liu, H., Li, J.: Lie symmetry analysis and exact solutions for the short pulse equation. Nonlinear Anal. Theory Methods Appl. **71**(5), 2126–2133 (2009)
13. Olver, P.J.: Applications of Lie Groups to Differential Equations, 2nd edn. G.T.M, Springer. New York (1993)

14. Röschel, O.: Die Geometrie des Galileischen raumes. Habilitationsschrift, Leoben (1984)
15. Yaglom, I.M.A.: Simple Non-Euclidean Geometry and Its Physical Basis. Springer, New York (1979)
16. Yoon, D.W.: Inelastic flows of curves according to equiform in Galilean space. J. Chungcheong Math. Soc. **24**(4), 665–673 (2011)
17. Yıldız, Ö.G., Tosun, M.: A note on evolution of curves in the Minkowski spaces. Adv. Appl. Clifford Algebras **27**(3), 2873–2884 (2017)
18. Yüzbaşı, Z.K., Anco, S.C.: Elastic null curve flows, nonlinear C-integrable systems, and geometric realization of Cole-Hopf transformations. arXiv preprint arXiv:1709.08234 (2017)
19. Yüzbaşı, Z.K., Bektaş, M.: A note on inextensible flows of partially and pseudo null curves in E_1^4. Prespacetime J. **7**(5), 818–827 (2016)
20. Yüzbaşı, Z.K., Aslan, E.C., Inc, M.: Lie symmetry analysis and exact solutions of Tzitzeica surfaces PDE in Galilean space. J. Adv. Phys. **7**(1), 88–91 (2018)

Index

© Springer Nature Switzerland AG 2020
J. A. T. Machado et al. (eds.), *Numerical Solutions of Realistic Nonlinear Phenomena*, Nonlinear Systems and Complexity 31, https://doi.org/10.1007/978-3-030-37141-8

Printed by Printforce, the Netherlands